This book

Ways of Learning and Knowing
The Epistemology of Education

is available as a free PDF download from the publisher's website, www.livingcontrolsystems.com, as well as the free online libraries www.archive.org and www.z-lib.org, which will help ensure that this book and others on the subject of **Perceptual Control Theory, PCT,** will be available to students for many decades to come.

File name: LearningKnowingPetrie2012.pdf

The file is password protected. Changes are not allowed. Printing at high resolution and content copying are allowed. Before you print, check the modest price from your favorite Internet bookstore. For related books and papers, search *Perceptual Control Theory*

For drop ship volume orders, mix and match, contact the publisher.

For biographical information about Hugh Petrie, see the publisher's website.

Minor updates and this note added in 2020.

Ways of Learning and Knowing

THE EPISTEMOLOGY OF EDUCATION

Hugh G. Petrie

Living Control Systems Publishing
Menlo Park, CA

Copyright © 2012 by Hugh G. Petrie

All rights reserved.

Library of Congress Control Number: 2012941574

Publishers Cataloging in Publication
Petrie, Hugh G. 1937 –
Ways of learning and knowing : The epistemology of education.
   xv, 356 p. : illustrated : bibliography: index; 24 cm.

   978-1-938090-06-6 (softcover, perfect binding)
   978-1-938090-07-3 (hardcover, case binding)

   1. Learning. 2. Schooling.
   3. Education. 4. Evolutionary epistemology
   5. Perceptual control theory. 6. Philosophy of Science
   I. Title.

LB1060.P   2012   370.15'23

∞ The paper used in this book meets all ANSI
   standards for archival quality paper.

On the back cover: Picture of Hugh Petrie taken in 2011.

Contents

Foreword. . . . . *vii*
Preface . . . . *xi*
Blind Variation and Selective Retention
 —An Intellectual Autobiography. *1*
Why Has Learning Theory Failed
 To Teach Us How To Learn?. *21*
Theories are Tested by Observing the Facts
 —Or Are They? *31*
Action, Perception, and Education. *57*
Can Education Find Its Lost Objectives
 Under the Street Lamp of Behaviorism? *77*
Do You See What I See?
 The Epistemology of Interdisciplinary Inquiry *91*
Metaphorical Models of Mastery:
 Or, How to Learn to Do the Problems at the
 End of the Chapter of the Physics Textbook *107*
A Rule by Any Other Name is a Control System. *121*
Evolutionary Rationality: Or Can Learning Theory
 Survive in the Jungle of Conceptual Change?. *143*
Metaphor and Learning *165*
Against "Objective" Tests: A Note on the
 Epistemology Underlying Current Testing Dogma *207*
Testing for Critical Thinking. *237*
Interdisciplinary Education: Are We Faced With
 Insurmountable Opportunities? *259*
Knowledge, Practice, and Judgement. *301*
A New Paradigm for Practical Research *319*
Purpose, Context, and Synthesis:
 Can We Avoid Relativism? *341*

Foreword

I spent 28 years in the College of Education of the University of Illinois. Hugh Petrie spent 10 years in the same college in a different department. Although our tenures at Illinois overlapped by two years, I cannot recall meeting Hugh at Illinois.

But even if we had met, I don't think we would have had much in common at that time. I was interested in bilingual education and first and second language acquisition. Hugh was an educational philosopher. Had we met, Hugh might have told me of his interests in the origin of knowledge and the process of learning. He might have told me how the work of Donald T. Campbell and William T. Powers had given him crucial insights into these fundamental educational issues. But as a young assistant professor, I probably would not have been much impressed, concerned as I was with getting my dissertation studies published and launching a research program that would secure my tenure at Illinois.

It was only several years later that I took the time to examine some of these basic issues in education and psychology. And, curiously, doing so also led me to the work of Campbell and Powers. Their work, together with the long chats I had with Don Campbell at Lehigh University and Bill Powers in Chicago, completely changed my perspective regarding how knowledge growth and learning is possible and why it is that people do the things they do. Such was their impact, that between 1989 and 2000 I focused almost all of my professional energy into the preparation of two books, *Without Miracles* and *The Things We Do*. Neither book would have been possible without knowledge of the work of Campbell and Powers.

I did meet Hugh at the April 1995 annual meeting of the American Educational Research Association, where I participated in a session with both Hugh and Bill Powers. Meeting Hugh motivated me to read his book, *The Dilemma of Enquiry and Learning. Dilemma* revealed to me how Meno's dilemma concerning inquiry can be resolved by seeing knowledge growth as a process involving adaptation, with Powers' perceptual control theory relevant to the elaboration and refinement of existing conceptual schemes and Campbell's blind variation and selective retention relevant to the creation of new conceptual schemes.

But Hugh's extensive writings deal with far more than basic questions of knowledge growth and learning. Over four decades Hugh has carefully considered and analyzed a broad array of issues in education, a range of topics that can be best be appreciated by reading his "intellectual autobiography" included as the first paper in this volume. I was frankly unaware of the range of his interests and expertise until seeing these sixteen published papers brought together in this volume.

For most of his career, Hugh was way ahead of his time. His papers in this volume still are. The role of the evolutionary process of blind variation and selective retention in all knowledge processes and the understanding of behavior as the control of perception are still mostly unknown in mainstream educational research, theory and philosophy. These perspectives, combined with Hugh's analytical skills and accessible writing, lead to some radical (and radically useful) implications for our understanding of the process of knowledge growth and the practice of education.

Gary Cziko
Professor Emeritus, University of Illinois
author of *Without Miracles* and *The Things We Do*
Urbana, Illinois, 2012

Preface

I have spent my entire professional life as a philosopher, philosopher of education, and educational administrator fascinated by the questions of how we learn and how we know what we learn. My attempt at putting my views on these subjects into a coherent whole is my book, *The Dilemma of Enquiry and Learning* (1981) Chicago: University of Chicago Press; revised and expanded (2011) Menlo Park, CA: Living Control Systems Publishing. However, as Dag Forssell, the editor of Living Control Systems Publishing, was putting out the revision of that book, he commented that there were quite a few of my articles and book chapters over the years that had both prefigured the book and later expanded in more detail on a number of its themes. He suggested that we put together an anthology of those articles and chapters and this volume is the result.

This anthology begins with my intellectual autobiography where I trace my path from a small town high school student through my higher education and on to my first position as a philosopher at Northwestern University. It was there that I met the two most important influences in my professional life. Donald Campbell taught me to appreciate and embrace evolutionary epistemology and William T. Powers taught me the revolutionary psychological theory that behavior is the control of perception, not the other way around. Those two influences profoundly permeated the rest of my professional life.

Even before these two influences had fully informed my thinking, I wrote *Why has Learning Theory Failed to Teach us How to Learn*. I there criticized behavioral psychologists for being unable to communicate their stimulus-response theories to educators who firmly believe that we do act to achieve our goals—an insight later beautifully explicated with the aid of perceptual control theory.

In *Theories are Tested by Observing the Facts—Or Are They?* I argued for the then emerging thesis of the theory-dependency of observation which has since become a staple of epistemology. I was also already anticipating the perceptual control theory insight that the same action, e.g., driving to work, or getting the food in a rat's puzzle box, could be accomplished in an indefinite number of ways with an indefinite array of muscle movements, etc. Behavior is the control of perception. We try to maintain our perceptions in the state we want to see and this can be accomplished in a dizzying array of environments.

Action, Perception, and Education is my first short attempt to explain perceptual control theory to educators. The main problem with any abbreviated attempt to explain perceptual control theory is that, as Kuhn has pointed out so well in The Structure of Scientific Revolutions, people resist scientific revolutions as hard as they can and stubbornly continue to try to explain the new paradigm in terms of their old familiar paradigms. So it is with perceptual control theory. One really has to study it at length with an open mind in order to appreciate its real revolutionary appeal.

In many of my writings, I had a penchant for catchy titles. This is evident in *Can Education Find Its Lost Objectives Under the Street Lamp of Behaviorism*. The continuing critique of behaviorism to be found there is part of the groundwork for looking elsewhere for a coherent theory of human behavior. Since behaviorism is as totally incoherent as its critics argue, it should be easier to understand and accept perceptual control theory.

As I was thinking through the implications of evolutionary epistemology and perceptual control theory, I was privileged to participate in an interdisciplinary faculty seminar at the University of Illinois examining the role of the social sciences and humanities in an engineering curriculum. I found a number of my ideas, especially the theory-dependency of observation to be clearly relevant to my experience in that seminar and *Do You See What I See? The Epistemology of Interdisciplinary Inquiry* was the result. In perceptual control theory the test for whether someone is perceiving something is to introduce a disturbance and see if it is counteracted. Thus there is a test for when someone has learned the observational and theoretical categories of a new discipline.

Given that observation and understanding are dependent on a given conceptual scheme, how is it that we can ever learn anything new; for we must always start with what we have? In *Metaphorical Models of Mastery: Or, How to Learn to Do the Problems at the End of the Chapter of the Physics Textbook* I sketched an early answer to the complete treatment I gave in *The Dilemma of Enquiry and Learning*. Metaphors and concrete examples allow us to bridge the gap between what we currently believe and the new material we are to learn.

In *A Rule by Any Other Name is a Control System* I examined the ubiquitous concept of "rules" in a variety of psychological theorizing. I argued that even behavior in accordance with so-called "descriptive" rules presupposes norms and require judgments as to the appropriateness of the norm. I then showed how an analysis of rule-following behavior using perceptual control theory meets all of the criteria for norm-regarding behavior in a completely transparent way.

Evolutionary Rationality: Or Can Learning Theory Survive in the Jungle of Conceptual Change? prefigured the use of evolutionary epistemology to account for conceptual change in my book, *The Dilemma of Enquiry and Learning*. I used the central idea of blind (not random) variation and selective retention to show how concepts can rationally change. This change occurs both in the growth of knowledge generally and in the growth of knowledge for individual students.

The *Metaphor and Learning* chapter by Petrie and Oshlag is a revision of the original chapter by Petrie which appeared in A. Ortony (1979). (Editor). Metaphor and Thought, First Edition. Cambridge. Cambridge University Press. It is the most extended of my treatments of metaphor as the key bridge accounting for rational change between conceptual schemes in the growth of knowledge, both in science and for individuals. Thus metaphors are not only useful in the educational process, they are epistemologically necessary.

The use of traditional paper and pencil tests in education is as ubiquitous as it is mostly misguided. I argued this point at length in *Against Objective Tests: A Note on the Epistemology Underlying Current Testing Dogma*. I showed that the "objectivity" obtained through interpersonal agreement, e.g., machine scoring, both limits what we can learn about what someone knows about a subject, and falsely leads one to believe that "subjective" tests, e.g., interviews, are somehow biased. Analyzing testing through the lens of perceptual control theory shows how interviews and other "subjective" tests often are the most reliable indicators of what someone knows. Introduce a disturbance to an hypothesized controlled variable and see if it is counteracted.

In my presidential address to the Philosophy of Education Society, *Testing for Critical Thinking*, I elaborated on the perceptual control theory analysis of finding out what someone knows and can do. Again the key is to introduce a disturbance to an hypothesized controlled variable and see if it is corrected. The doctoral oral is a paradigm example of how one can apply this notion and follow it up with additional probes and disturbances to determine if the candidate really can think critically.

In 1992 I updated my earlier work on interdisciplinary education in *Interdisciplinary Education: Are We Faced With Insurmountable Opportunities?* I considered the "disciplinary paradox"—the idea that the fragmentation of knowledge into disciplines calls for an interdisciplinary approach, but can only receive epistemic justification from the established disciplines. This paradox is a close relative to the Meno dilemma I dealt with in *The Dilemma of Enquiry and Learning*. And the solution is similar. Once one recognizes that knowledge is both theoretical and practical, attention to thought and action as justification allows one to avoid the paradox. Both disciplinary and interdisciplinary knowledge are justified because they allow us to pursue our human purposes in an ever-changing, but broadly stable, world.

By the time I wrote *Knowledge, Practice, and Judgment* I was increasingly utilizing my epistemological insights in the service of educational policy analysis. In this piece I criticized the notion that teachers should "apply" research to practice. The argument is the familiar one that the theories and categories of the researcher are largely incompatible with the concepts and perceptual categories of the teacher. The concept of professional judgment, on the contrary, allows for the fact that teachers can and do adapt their knowledge and action to constantly changing situations to further their goals.

I continued the emphasis on educational policy in *A New Paradigm for Practical Research*. I elaborated on how perceptual control theory with its insistence that behavior is the control of perceptions provides a real underlying model of how human action works. It provides a physically plausible explanation both for the consistency of outcomes of human action and the variability of means utilized to achieve those outcomes in a constantly changing environment. I urged that the then emerging concept of professional development schools provides the perfect real world laboratory for research using a perceptual control theory model of human behavior.

Finally, in *Purpose, Context, and Synthesis: Can We Avoid Relativism* I commented on several articles by evaluators worrying about the validity of evaluations that take into account the purpose of the evaluation and its context. Once again with a perceptual control theory model of evaluation, these concerns disappear. By sensing the various nuances of context, we are in effect comparing the actual context with our concept of that which is being evaluated and we need not know in advance what the context might be. We need only describe the extent to which the actual situation meets or fails to meet the reference concept in order to make warranted evaluative judgments.

Hugh Petrie
Tucson, Arizona, 2012

[2008]
Blind Variation and Selective Retention
—An Intellectual Autobiography

Over 50 years ago I sat in the living room of my home in a small town in Colorado where the Dean of Admissions of the California Institute of Technology was interviewing me for freshman admission to their engineering program. As we spoke, the dean said that many intelligent students in small town high schools gravitate toward science and mathematics courses, since these are often the only ones that are intrinsically interesting. Without excellent teachers in the humanities and social sciences, such students may not realize the intellectual attractions of those fields.

A "broad" undergraduate education

I was ultimately admitted to Caltech and MIT as well as to the University of Colorado. Because I was awarded a very generous scholarship to Colorado, I enrolled there, but, heeding the words of the Caltech dean, I naively thought I would expand my horizons by entering a joint five-year engineering and business program! It was the beginning of blind variation and selective retention although I could not have put those words to it at the time.

Fortunately, after my first two years at Colorado I discovered that I could also sit in on general honors courses, and eventually, I became the first engineering student at Colorado to enroll full-time in that program. In my first honors seminar, I read Descartes. "Cogito, ergo sum!" Wow! That dean was right; there was a whole new intellectual world in the humanities. By then, however, I was so far along in my applied mathematics and business studies that I decided simply to finish them rather than shifting at that time to my newfound love of philosophy. My advisors also told me that since analytic philosophy was the intellectual

---

Reproduced with permission of publisher from: Leonard J Waks (Ed), *Leaders in Philosophy of Education*, (2008) Sense Publishers, Rotterdam, pp 159 - 172.

fashion of the day, my work in mathematics would stand me in good stead. I received a Fulbright Fellowship to Manchester University to study mathematics and philosophy for a year and then I took up Woodrow Wilson and Danforth Fellowships to study philosophy at Stanford.

Linguistic analysis

As befitted the vogue in philosophy in the early 60's, I was thoroughly indoctrinated at Stanford into the reigning forms of linguistic analysis. However, I had always been interested in educational issues, and, in particular, in how we learn and come to know what we know. I gravitated to epistemology in my studies and I was particularly influenced by Israel Scheffler's *The Language of Education* (1960). Scheffler's later book, *Conditions of Knowledge* (1965) was copyrighted the same year as my doctoral dissertation, *Rote Learning and Learning With Understanding*, and the philosophical resonance between the two is really quite remarkable.

At Stanford, I also established a relationship with Larry Thomas, the senior philosopher of education at Stanford's School of Education. I was able to assist him in a couple of summer courses in philosophy of education (a bit more blind variation).

Evolutionary epistemology and Perceptual Control Theory

Following my doctoral work at Stanford, my first academic position was in the philosophy department at Northwestern University. There I made the acquaintance of Joe Park, the philosopher of education in the School of Education, who encouraged my early research and writing in philosophy of education, mostly an elaboration of the analytic epistemological themes drawn from my doctoral dissertation (Petrie, 1968, 1969, 1970).

It was also at Northwestern that the major influences on my intellectual development occurred. During my first year as an assistant professor, I was visited by Donald Campbell, the social psychologist, innovative social science methodologist, and "closet" philosopher of science. It is from Campbell's work in evolutionary epistemology that I have drawn the title, *Blind Variation and Selective Retention*, for my contribution to this volume. That phrase, as I will elaborate in what follows, sums up not only my intellectual autobiography, but also my views on how we come to know what we know.

Campbell was going on sabbatical during my first year at Northwestern, but he had heard that the philosophy department had hired an epistemologist and philosopher of science, and he wanted to ask me to co-teach a standard course he offered the next year when he returned. The course was entitled, "Knowledge Processes," and I said I would be happy to do so as long as my chair agreed. (Interestingly, it was this experience with Don Campbell that later encouraged me as a dean to encourage joint teaching experiences by my faculty, even if it didn't quite constitute a "regular" teaching load.)

It was during that course that I was first introduced to Thomas Kuhn (1962), Stephen Toulmin (1963), Karl Popper (1965), N. R. Hanson (1958), and, of course, to Donald Campbell. In the course, I read early drafts of his landmark *Evolutionary Epistemology* (1974). However, it took me awhile to selectively retain all the wonderful blind variations I was introduced to during that course. In fact, as a newly minted Ph.D. (does anyone know more than brand-new Ph.D.'s?), I was amazed that this well-known and highly respected full professor could be making so many elementary epistemological mistakes; mistakes that I had learned to refute during my graduate studies in analytic philosophy. So we had a number of robust discussions in the course about the theses Campbell was presenting. The students, quite naturally, loved the back and forth between the professors. After several months of Don Campbell's patient explanations of his position and questioning of my arguments, I began to think that maybe this full professor knew more than I had originally assumed. By the time we co-taught the course several times in the following years, I was beginning to see the outlines of how evolutionary epistemology might just be able to solve some of the continuing vexing philosophical questions of how we know and how we come to know. Over the years, I continued to keep in touch with Don Campbell and read all that he published on evolutionary epistemology.

The other major influence on my intellectual development also occurred at Northwestern, and, once again, was the result of blind variation and selective retention. The blind variation came from my attending a series of informal luncheon get-togethers organized by Don Campbell. At those luncheons, I made the acquaintance of William Powers. Powers was a true iconoclast. He earned his bachelor's degree in physics, and then enrolled in a doctoral program in psychology to pursue his interests in the connections between certain engineering concepts and human behavior. He left without finishing his degree in psychology in disgust with the reigning behaviorist ideology in psychology.

When I met Bill Powers, he was working as an engineer at a research facility at Northwestern and attempting to pull together his insights into human behavior into a book. At several of our luncheon meetings, he gave demonstrations of what he came to call Perceptual Control Theory. These demonstrations served both as striking refutations of stimulus-response psychology and as incredibly compelling illustrations of Perceptual Control Theory. (For the interested reader, some of these original demonstrations, others developed later, and a general introduction to Perceptual Control Theory can be accessed at www.livingcontrolsystems.com. Also see Powers (1998) for a basic introduction to Perceptual Control Theory.)

I was fascinated by this initial brief exposure to Powers' work and I determined to learn more about it and to try to give it a broader exposure. Consequently, I asked him to co-teach a graduate seminar with me on his work. Only about a half dozen Northwestern students signed up, but it was a mind-bending experience for all of us. Bill had us read draft chapters of the book he was working on, showed us many more demonstrations, and engaged us in the most exciting intellectual experience I had ever had. Those chapters later became his seminal book, *Behavior: the Control of Perception* (1973).

Linguistic analysis

My earliest work at Northwestern was still largely influenced by my doctoral training in analytic philosophy. Even in these writings, however, there were glimmers of the more full-blown emphasis on conceptual change, knowledge acquisition, Perceptual Control Theory, and a naturalized, evolutionary epistemology which came to dominate my later work. In *The Strategy Sense of 'Methodology'* (1968), I used the language of logical analysis to argue for the importance of the processes of obtaining knowledge and not just analyzing states of knowing or knowing how. In *Science and Metaphysics: A Wittgensteinian Interpretation* (1971a), I was already propounding the continuity of science and philosophy, as opposed to the linguistic analysts who held that philosophy was all about grammar. This lengthy book chapter used that paradigmatic linguistic philosopher, Wittgenstein, as a source for hints as to what I and others later came to call naturalized epistemology.

Don Campbell's influence was already apparent in another one of my early papers still couched primarily in the idiom of logical analysis. *A Dogma of Operationalism in the Social Sciences* (1971) argues that the behaviorists' beloved concepts of reliability and validity as exhibited in operational definitions are actually relative to what we take as an observation language. Contrary to the beliefs of most behaviorists, it is simply unsupported dogma to believe that there is some a priori set of observational terms, e.g., atoms of behavior, to which we can always unproblematically refer.

Naturalized evolutionary epistemology, conceptual change, and the theory-ladeness of observation in educational philosophy

In 1971 I moved to the College of Education at the University of Illinois at Urbana-Champaign as a philosopher of education. By then my work was beginning to reflect not only the influences of my time at Northwestern, but also my increasing interest in setting my work in educational contexts. Following my linguistic analysis phase at Northwestern, four major substantive themes emerged in my scholarly writing while at Illinois. Some of these were prefigured in my early work, but they only began to emerge full-blown after my move to Illinois. The first theme is the constellation of topics encompassed by a naturalized evolutionary epistemology, conceptual change, the theory-ladeness of observation, and critiques of behaviorism, especially as used in education. Second, my work in the epistemology of interdisciplinary inquiry also grew out of these topics. Third, my investigations into metaphor provide the key to understanding how conceptual change is possible and why metaphors are of such paramount importance to learning. Fourth, Perceptual Control Theory as an analysis of human behavior explains how entities control what happens to them and illustrates the relationships between actions and goals, perceptions and actions, and perceptions and reality. Furthermore, Perceptual Control Theory does so within a single, testable concept of how living systems work.

The theory-ladeness of observation

Why Has Learning Theory Failed to Teach Us How to Learn (1968) applies the relativity of observational languages to stimulus-response learning theorists on the one hand and educational practitioners on the other. The former use behavioral observational categories and the latter mentalistic action categories. The two camps pass each other in the night. *Theories Are Tested by Observing the Facts: Or Are They?* (1972), expands considerably on the learning theory article. In particular I argue there that non-behavioral approaches in educational research cannot simply be ignored and that an eclectic "functionalism" in educational research is bound to try to compare apples and oranges and, hence, end up being incoherent. Only fully articulated Kuhnian paradigms can fruitfully be compared in terms of educational research. *Can Education Find Its Lost Objectives Under the Street Lamp of Behaviorism?* (1975), applies the lessons of *Dogma* (1971b) to a thoroughgoing critique of the educational policy of utilizing behavioral objectives as the panacea for all, or almost all, educational ills.

Interdisciplinary inquiry

Perhaps my best-known, and most reprinted, article relies heavily upon the theory-ladeness of observation. Early in my career at Illinois, I was invited to join a group of engineers, natural scientists, social scientists, and humanists who were funded by the Sloan Foundation to explore, in an interdisciplinary way, the role of the social sciences and humanities in an engineering curriculum. The method was to hold interdisciplinary seminars of all the faculty participants. Each seminar was led by an expert in a different discipline, engineering, humanities, social science. The faculty member of the moment attempted to answer the question, "How does my discipline view the world?" If ever there was a real-life exploration of the theory-ladeness of observation, this was it! It formed the impetus for my work in the epistemology of interdisciplinary inquiry and led to my most widely republished paper, *Do You See What I See? The Epistemology of Interdisciplinary Inquiry* (1976a). In this paper, I argued that truly interdisciplinary inquiry can proceed only if there is at least a rudimentary understanding of the observational categories, and, hence, theory, of the various disciplines involved. This explains why interdisciplinary work is so hard. You almost have to acquire a new discipline. A later reflection on interdisciplinary education can be found in *Interdisciplinary Education: Are We Faced with Insurmountable Opportunities?* (1992a).

Metaphor

So, are we faced with insurmountable difficulties because we have to learn the concepts and observational categories of the new discipline? Not quite. From the seminars, I learned that well-chosen and elaborated metaphors can at least begin to provide the insights necessary to understand one's partners in an interdisciplinary effort. This realization was strengthened by my Illinois colleague, Andrew Ortony. His extensive work on metaphor (1975, 1979, 1993) Is a gold mine for the student who wishes to pursue this line of work. I was also influenced by my truly remarkable graduate students at the time who helped me refine my thinking on metaphors and conceptual change. It was to try to solve the problem of how we learn new conceptual schemes that my work on metaphor emerged.

Of course, the interesting implication for education is that students learning a new discipline are in the same position as the participants were in the interdisciplinary seminars. They all need to learn the theory-laden observational categories of the discipline without the benefit of any a priori neutral set of observations. Students do, however, have a teacher and good teachers are able to use well-chosen metaphors to help bridge the gap between the common sense observational categories of the student and the observational categories to be learned in the discipline. I argue these points in *Metaphorical Models of Mastery: Or, How to Learn to do the Problems at the End of the Chapter of the Physics Textbook* (1976b). I believe that Kuhn's (1974) notion of "exemplars" i.e., exemplary problem solutions, is part of what allows the metaphors to be successful. I also suggest here that the scientist involved in conceptual change at the frontiers of the discipline is, in many ways, analogous to the student. Both need to try out new observational categories with the help of metaphors. The student has the teacher to help weed out bad interpretations by guiding the student through the new field with demonstrations, lab exercises, homework, and the like. The scientist has "nature" as teacher. Experiments are performed and they help weed out incorrect predictions, hypotheses, and observational categories. My most detailed account of how metaphors work for both the student learning something new and for the scientist on the frontiers of knowledge can be found in *Metaphor and Learning* (1979a) and in the revision of that book chapter with Rebecca Oshlag (1993).

Naturalized evolutionary epistemology

Returning to my first theme at Illinois, I eventually came to see that a naturalized evolutionary epistemology was necessary to encompass all of these insights—the theory-dependency of observation, the growth of knowledge, conceptual change, metaphors, and critiques of behaviorism. And the key slogan for that epistemology provides the title to this chapter, *Blind Variation and Selective Retention*. I adopt this phrase from Don Campbell's brilliant piece, *Evolutionary Epistemology* (1974). There is no way that I can fully elucidate this idea in the short space I have available here, nor can I begin to deal with the numerous "standard" objections to a variation and retention view of evolution, whether biological or conceptual. Campbell does a wonderful job, and I devote considerable space to this topic in my book, *The Dilemma of Enquiry and Learning* (1981). For now let me simply say that "blind" does *not* mean "random." Rather it means that although *what* is being varied, e.g., concepts, theories, even organisms, does not know beforehand what will be encountered, these variants, because they have survived thus far, already contain a good deal of at least partial wisdom about the environment. We don't start from scratch varying "atoms." Furthermore, selection need not involve the complete elimination of some variants. There are "vicarious" selection mechanisms at work too, e.g., generally accepted common sense theories, other scientific theories not at the moment subject to examination, long-standing common sense observational categories. None of these are a priori infallible, and each may be questioned in its turn, but they at least have worked tolerably well up until now. Finally, although we can never have direct access to "reality as it really is," there is a role for a reality that forms the basis against which we test, change, and test again our representations of it. I am a realist.

My first, short account of how evolutionary epistemology can deal with conceptual change is to be found in *Evolutionary Rationality: Or Can Learning Theory Survive in the Jungle of Conceptual Change?* (1977a). In this paper I argue that given the theory-dependency of observation a philosophical concern for truth cannot be taken simply as some sort of direct correspondence between our observations and conceptual schemes on the one hand and "reality" on the other. Rather, we must consider how our observational categories and conceptual schemes as a whole allow us to deal with the world in terms of all of the human purposes, social and individual, that we have. Thus, although I read very little Dewey or James or other pragmatists, I believe that I echo some pragmatic themes in my work. This relation to pragmatism becomes even more evident in *Science and Scientists, Technology and Technologists, and the Rest of Us* (1977b). In this book chapter I explicitly consider the relationship between evolutionary epistemology and pragmatism. Specifically, I argue that evolutionary epistemology can assist pragmatism with several of the traditional challenges to its justification of science. Evolutionary epistemology can help locate sources of values in science while still allowing for the "objectivity" of science and technology that is found in the disciplinary aspects of science. It can, by taking an appropriately long-term and expansive view of the development of science help defend the value of science and technology from isolated counterexamples in which science and technology have not led to humane results. Finally, evolutionary epistemology can help pragmatism deal with the objection that inappropriate social power distributions might capture the social arrangements of the disciplines. This can happen here and there, e.g., in "scientific" objections to global warming funded by industry. But the fact that the scientific disciplines often find other social arrangements to further their work, e.g. universities, "green" organizations, suggests that the discipline will continue to evolve and answer our basic human purposes.

Perceptual Control Theory

Let me turn now to the fourth major theme in my work at Illinois. Just what is "Perceptual Control Theory?" Unfortunately, I can hardly explain it adequately in the brief space available here. I again refer you to the web site mentioned above where you can explore a number of different introductions to the theory, along with on-line demonstrations. The most important insight is that human beings employ negative feed back systems to control their *inputs*, i.e., their *perceptions*, rather than their outputs, i.e., their behaviors. Behavior is used to control our perceptions of our environment and these perceptions are compared with what we want to see, i.e., our purposes and intentions in acting. We then vary our outputs, not with any sort of detailed "plan", but almost automatically. These outputs affect the world which in turn affects out perceptions, bringing them closer to what we want to see in the case of well-adjusted control systems. Think of driving a car. We don't calculate which way to turn the wheel when the road turns or a crosswind takes us out of the lane, we just automatically turn the wheel until we perceive the car where we want it to be in conjunction with the road. [1]

My first attempt at introducing Perceptual Control Theory (PCT) to educational audiences was *Action, Perception, and Education* (1974).

---

[1] Those familiar with the educational literature will recognize that William Glasser has written extensively in education utilizing a concept he calls "control theory." Although there are superficial resemblances to Powers' Perceptual Control Theory, Glasser completely fails to appreciate that what is controlled are perceptions, not actions or behaviors. This renders Glasser's version of control theory no more insightful than most cognitivist theories in psychology.

It fell stillborn from the press. I followed this attempt to present the whole of PCT in one article with a number of more pointed educational implications. In *A Rule by Any Other Name is a Control System* (1976c) I argue that any number of problems with the analysis of rule-following in psychology can be solved by treating rule-following as the operation of control systems rather than as some mysterious and complicated associationist view of habits.

In *Against "Objective" Tests: A Note on the Epistemology Underlying Current Testing Dogma* (1979) I show how "objective" tests as they are understood in the evaluation literature are sorely limited in how much they can actually tell about the competence and knowledge of those who are being tested. On the other hand, "subjective" tests are much more nuanced and capable of revealing the depth of understanding of the person being tested. All of this follows from a principle of PCT, "the test for the controlled variable." The test for the controlled variable is a method for finding out just what perceptions someone else is actually controlling for with their behavior. The test proceeds by introducing what would be disturbances to the hypothesized variable that one thinks the person is controlling and seeing what they do to counteract those disturbances. Thus, it is no accident that one of the most intellectually challenging tests we have, the Ph.D. oral, allows for the examiners to vary their questions to explore just what the candidate really has in mind. Even doctoral prelims are typically of an essay variety where the candidate can counteract the disturbance introduced by the questions. We certainly do not give Ph.D. candidates "objective" true-false or multiple choice tests. My most elaborate exposition of how PCT helps us ground testing is to be found in my Philosophy of Education Society Presidential Address, *Testing for Critical Thinking* (1986).

In *Program Evaluation as an Adaptive System*, (1982) I apply the notions of PCT and adaptive systems to argue that in order for program evaluation to be integrated into an institution's structure rather than resisted by it, both the evaluation scheme and the institution must be viewed as adaptive systems which control their perceptions. I also suggest in *Purpose, Context, and Synthesis: Can We Avoid Relativism?* (1995b) that evaluation specialists who insist that evaluation research must be tied to the context and the purposes of the evaluation can, nevertheless, reach warranted conclusions. They do not need to retreat to positivism or be branded as relativists. Indeed, human beings, conceived of as control systems, are able to achieve consistent results in a constantly changing environment. Thus in evaluating how they do that, we *must*, as evaluators, look at both what the actors are trying to achieve and at how the context in which they are doing this is changing.

The Dilemma of Enquiry and Learning

Clearly the most comprehensive and detailed analysis I give of the various themes encompassed in my philosophical work is to be found in my book, *The Dilemma of Enquiry and Learning* (1981). In this work, I take Plato's Meno dilemma seriously. The dilemma says that we can neither inquire into anything which we know nor into anything which we do not know. For if we already know something, we have no need to inquire, but if we do not know something, we cannot inquire, for we would not know where to begin nor when we had reached knowledge of what we do not know. In short, the Meno dilemma seems to pose the Kantian question, "How are inquiry and learning possible?"

In brief, my solution to the Meno dilemma (after extensive exposition and argument) is that we must step between the horns of the dilemma by giving both of them their due. One of the major preconditions for stepping between the horns is to argue that we must focus on knowledge *processes* rather than knowledge *structures*. "Knowing" and "learning" are the fundamental notions rather than "knowledge" and "what is learned." If we focus on knowledge process, we can see that even what I called the "old knowledge" horn of the dilemma is really quite sharp. Just because we "know" something in the sense of having acquired a knowledge structure it does not follow that we automatically know how to apply that structure in a constantly changing environment to achieve consistent results. Recall the example of driving a car. Almost everyone already "knows" how to drive. Yet each time we are on the road, even on our well-worn route to the office or the grocery store, we are faced with different circumstances with which we must cope in order to get where we are going. The behaviorist and even traditional cognitivist psychological approaches to explaining our continuing successes in such situations face insurmountable difficulties.

In the book I call the knowledge process that accounts for our ability to utilize existing conceptual frameworks in changing circumstances, "assimilation," and while the term is similar to Piaget's use of the same language, I do not give it a Piagetian elaboration. Rather, I present Perceptual Control Theory and show how it transparently shows how conceptual structures conceived as perceptual control systems and hierarchies of control systems explain our ability to achieve consistent results in very different environments.

Of course, the "new knowledge" horn of the dilemma is very sharp as well. Occasionally, we really do need to radically change our conceptual schemes, whether we be a scientist on the frontiers of knowledge or a student just learning a brand new discipline that is incompatible with the student's existing beliefs. I call the knowledge process that accounts for radical conceptual change, "accommodation," although again the concept is not the Piagetian one. I argue that the blind variation and selective retention mechanism elaborated in a naturalized evolutionary epistemology is what is needed to account for successful processes leading to new knowledge structures.

The way between the horns of the dilemma lies in recognizing a reflective equilibrium between assimilation and accommodation. In dealing with the world we almost always try to assimilate new situations by means of our existing knowledge processes. However, if we continuously fail to be successful, we may need to try new structures. These new trials are best understood as metaphors that have to be tested against the world through whatever observational categories we happen to be using. Gradually, both metaphors and observations are brought into a kind of equilibrium, at least for the moment.

Educationally, I argue that we seem not to recognize the need for *both* assimilation and accommodation. Still less are we aware of when one ought to be stressed and when the other. In any educational situation we need to carefully analyze whether we are trying to get a student to refine an existing knowledge process and when we are trying to get the student to acquire new knowledge processes. We must always be striving for a reflective equilibrium between assimilation and accommodation in both our classroom practice and our educational policy making.

The book brings together in one place almost all of my thinking about educational epistemology. It utilizes themes from conceptual change, the centrality of metaphor, a focus on knowledge processes, the new psychology of Perceptual Control Theory, and a naturalized evolutionary epistemology.

Transitions

During the year's in which I was writing *Dilemma*, I was also undertaking a number of new blind variations in both my personal and professional lives. I divorced and remarried. I have now been married for 28 years to my wife, Carol Hodges. During this period, some of my writing on accountability and evaluation as well as my work in interdisciplinarity apparently came to the attention of the higher administration at Illinois. I was asked to take over as the director of the campus-wide program evaluation system at the university. Since it was just about my turn to assume a term as chair of my department, I blindly decided that the university-wide administrative position would be more interesting and probably less challenging than departmental politics. I was certainly wrong about the latter assumption. However, the opportunity to utilize my expertise in interdisciplinary inquiry to assist a blue ribbon campus committee of professors from different disciplines pass evaluative judgments on their colleagues' departments was one of the high points of my administrative career. And all of this was going on while I was writing my book!

The next blind variation came with our decision to move to Buffalo. Since Carol was a Ph.D. graduate of the University of Illinois, she was only able to teach there for several years on soft money and we were constantly on the lookout for a place where we could both obtain academic positions. By 1981 I had completed my stint as a campus-level administrator, but there were almost no openings for a philosopher of education, at least at institutions that also were looking for a reading and elementary education professor. So I started looking for administrative positions and in 1981 Carol accepted a faculty position at State University College at Buffalo and I accepted the deanship of the school of education at the State University of New York at Buffalo—both SUNY institutions, but separate.

A philosopher dean

This began my 16 year tour of duty as dean, followed by two years back as a professor before retirement. During my tenure as a dean, my professional focus turned largely to educational policy issues, although still strongly influenced by my philosophical beliefs. There were several strands to this focus. In 1987 a number of colleagues and I founded the journal, *Educational Policy*. A number of the "themed" issues from the journal were fleshed out and became edited books (Weis, L. et al. *Crisis in Teaching: Perspectives on Current Reforms*. 1989a), (Weis, L. et al. *Dropouts from School: Issues, Dilemmas and Solutions*. 1989b), (Altbach, P.G. et al. *Textbooks in American Society: Politics, Policy, and Pedagogy*. 1991a), (*Weis, L. et al. Critical Perspectives on Early Childhood Education*. 1991b), (Petrie, H.G. *Professionalization, Partnership, and Power*. 1995). A second focus emerged from my role as one of the founders of the institutional educational reform movement known as the Holmes Group (see the Holmes trilogy, *Tomorrow's Teachers* (1986), *Tomorrow's Schools* (1990), and *Tomorrow's Schools of Education* (1995). Although the Holmes Group as an organization is no more, the ideas it propounded have had a significant impact on teacher education. Extended preparation programs, a strong liberal arts education, a rejuvenation and strengthening of professional training, the concept of professional development schools as a joint project of real schools and schools of education, an emphasis on more practice-oriented research by education professors in research universities—all are now part of the educational landscape in one form or another.

I wrote on extended preparation and the liberal arts in teacher education (1987a, 1987b), strengthening professional preparation (1990), and professional development schools (1995a). I also continued to utilize my interests in educational epistemology in my policy writings. In *Knowledge, Practice, and Judgment* (1992b) I argued that we must substitute a notion of teacher judgment for that of "applying" research to practice. The latter depends for its justification on discredited views of knowledge processes, while the former takes full account of the view of knowledge processes I describe in *Dilemma*. Finally, in *From 'My Work' to 'Our Work'*, (1998) I reflected on my experiences as a dean in trying to encourage changes to the faculty culture in schools of education in research universities. Instead of the faculty viewing themselves as more or less independent intellectuals who happen to have a mailing address and Email account at a university, I tried over my years as a dean to encourage more collaborative teaching, research and outreach activities with the rest of the education profession—a shift from "my work" to "our work." At best, I had modest success.

Conclusion

Nevertheless, as I suggested in my valedictory address to the last group of students who graduated under my deanship, our efforts in the academy, whether teaching, research, service, or administration, are all a work in progress. As an education profession we refine our knowledge here and there and occasionally, blindly stumble across something quite new. Once in awhile, those blind variations are selectively retained and our profession lurches forward. The best each of us can do is make our own individual contributions and hope that some will "stick." That is what I have tried to do since my first encounter with that dean of admissions from California Institute of Technology over 50 years ago. I have undertaken one blind variation after another, starting with "broadening" my undergraduate education to include business as well as engineering. As it turns out my undergraduate business degree stood me in good stead as a dean 30 years later. I stumbled onto Don Campbell and Bill Powers and they changed my intellectual life in the most profound ways. I participated in an interdisciplinary seminar and became fascinated with the topic. I became a campus level administrator to avoid being a chair and was then able to find employment as a dean so that my wife and I could pursue joint careers in education. As a dean I put my philosophical background to work in furthering the cause of educational reform. I varied a lot of things, the outcomes of which I certainly could not have predicted in advance. But I used my knowledge and experience and values to select and retain what I hope were the best of those variants. I can only hope that others will carry on the work in progress that is educational philosophy.

Hugh Petrie's favorite works

Campbell, D.T. (1974). Evolutionary Epistemology. In P.A. Schilpp (Ed.) *The Philosophy of Karl Popper. The Library of Living Philosophers, vol. 14.* LaSalle, IL: Open Court.

Hansen, N.R. (1958). *Patterns of Discovery.* Cambridge: Cambridge University.

Kuhn, T. (1970). *The Structure of Scientific Revolutions.* Enlarged Edition. Chicago: University of Chicago Press.

Petrie, H.G. (February 1976). Do You See What I See? The Epistemology of Interdisciplinary Inquiry *Educational Researcher,* 5(2), 9-15.

Petrie, H.G. (1981) *The Dilemma of Enquiry and Learning.* Chicago: University of Chicago Press. [1]

Popper, K. (1965). Conjectures and Refutations: The Growth of Scientific Knowledge. 2nd edition. New York: Basic Books.

Powers, W.T. (1973). *Behavior: The Control of Perception.* Chicago: Aldine.
—Second edition (2005), revised and expanded, Bloomfield, NJ: Benchmark Publications

Powers, W.T. (1998). *Making Sense of Behavior: The Meaning of Control.* Bloomfield, NJ: Benchmark Publications.

Toulmin, S. (1963). *Foresight and Understanding.* New York: Harper and Row.

References/Bibliography

Altbach, P.G., Kelly, G.P., Petrie, H.G., and Weis, L. (eds.) (1991). *Textbooks in American society: politics, policy, and pedagogy.* Albany, NY: SUNY Press.

Campbell, D.T. (1974). Evolutionary Epistemology. In P.A. Schilpp (Ed.) *The philosophy of Karl Popper. the library of living philosophers, vol. 14.* LaSalle, IL: Open Court.

Hansen, N.R. (1958). *Patterns of discovery.* Cambridge: Cambridge University Press.

Holmes Group, (1986). *Tomorrow's teachers: A report of the Holmes Group.* East Lansing, MI: Author.

Holmes Group, (1990). *Tomorrow's schools: A report of the Holmes Group.* East Lansing, MI: Author.

Holmes Group, (1995). *Tomorrow's schools of education: A report of the Holmes Group.* East Lansing, MI: Author.

Kuhn, T. (1970). *The structure of scientific revolutions.* Enlarged Edition. Chicago: University of Chicago Press.

Kuhn, T. (1974). Second thoughts on paradigms. In F. Suppe (ed.). *The structure of scientific theories.* Urbana, IL: University of Illinois Press.

---

1 Revised and expanded (2011) Menlo Park, CA: Living Control Systems Publishing

Ortony, A. (1975). Why metaphors are necessary and not just nice. *Educational Theory, 25,* 45-53.

Ortony, A. (1979). (Editor). *Metaphor and thought,* First edition. Cambridge: Cambridge University Press.

Ortony, A. (1993). (Editor). *Metaphor and thought,* Second edition. Cambridge: Cambridge University Press.

Petrie, H.G. (1965). *Rote learning and learning with understanding* (Ph.D. Dissertation) Stanford University, 1965.[2]

Petrie, H.G. (1968). The strategy sense of 'methodology'. *Philosophy of Science, 35*(September), 248-257.

Petrie, H.G. (1969). Why has learning theory failed to teach us how to learn. *Proceedings of the Philosophy of Education Society, 1968.* 163-170.

—Reprinted in Broudy, H. S., Ennis, R. H., and Krimerman, L.I. (eds.), (1973). *Philosophy of educational research.* New York: John Wiley and Sons. 121-138.

Petrie, H.G. (1970). Learning with understanding. In Martin, J.R. (ed.), *Readings in the philosophy of education: a study of curriculum.* Boston: Allyn-Bacon. 106-121.

Petrie, H.G. (1971a). Science and metaphysics: A Wittgensteinian interpretation. In Klemke, E.(ed.), *Essays on Wittgenstein.* Urbana: University of Illinois Press. 138-169.

Petrie, H.G. (1971b). A dogma of operationalism in the social sciences. *Philosophy of Social Science, 1.* 145-160.

Petrie, H.G. (1972). Theories are tested by observing the facts: or are they? *Philosophical redirection of educational research, National Society for the Study of Education yearbook,* 71st Yearbook. Chicago: University of Chicago Press. 47-73.

Petrie, H.G. (1974). Action, perception, and education. *Educational Theory, 24,* 33-45.

Petrie, H.G. (1975). Can education find its lost objectives under the street lamp of behaviorism? In Smith, R. (ed.) *Regaining educational leadership: essays critical of PBTE/CBTE.* New York: Wiley. 64-74.

Petrie, H.G. (1976a). Do you see what I see? the epistemology of interdisciplinary inquiry. *Educational Researcher, 5*(2), 9-15.

—Republished in (1976a) *The Journal of Aesthetic Education, 10*(1). 29-43.

—Reprinted in (1986).Chubin, D. E., Porter, A. L., Rossini, F. A., and Connolly, T. (eds.), *Interdisciplinary analysis and research.* Mt. Airy, Maryland: Lomand. 115-130.

---

2 Posted under "About_Hugh_Petrie" at publisher's website.

▌ = Reprinted in this anthology

Petrie, H.G. (1976b). Metaphorical models of mastery: or, how to learn to do the problems at the end of the chapter of the physics textbook. In Cohen, R. S., Hooker, C. A., Michalos, A. C., and Van Evra, J. W. (eds.), *PSA 1974*, Dordrecht-Holland: D. Reidel. 301-312.

Petrie, H.G. (1976c). A Rule by Any Other Name is a Control System. *Cybernetics Forum, VIII*, (Fall/Winter). 103-114.

Petrie, H.G. (1977a). Evolutionary rationality: or can learning theory survive in the jungle of conceptual change? *Proceedings of the Philosophy of Education Society, 1976*. 117-132.

Petrie, H.G. (1977b). Science and scientists, technology and technologists, and the rest of us. In LaBrecque, R. and Crockenberg, V. (eds.), *Culture as education*. Dubuque, Iowa: Kendall-Hunt. 79-110.

Petrie, H.G. (1979a). Metaphor and learning. In Ortony, A. (ed.), *Metaphor and thought*, First Edition. (Cambridge: Cambridge University Press. 438- 461. [3]

Petrie, H.G. (1979b). Against 'objective' tests: a note on the epistemology underlying current testing dogma. In Ozer, M. N. (ed.), *A cybernetic approach to the assessment of children: toward a more humane use of human beings*. Boulder, Colorado: Westview Press. 117-150.

Petrie, H.G. (1981). *The dilemma of enquiry and learning.* Chicago: University of Chicago Press.[4]

Petrie, H.G. (1982). Program evaluation as an adaptive system. In Wilson, R. (ed.), *New directions for higher education: designing academic program reviews, no. 37*. San Francisco: Jossey-Bass. 17-29.

Petrie, H.G. (1986). Testing for critical thinking (Presidential Address). *Proceedings of the Philosophy of Education Society 1985*. 3-20.

Petrie, H.G., (1987a). The liberal arts and sciences in the teacher education curriculum, In Carbone, M.J. and Wonsiewicz, A. (eds.), *Excellence in teacher education through the liberal arts*. Muhlenberg, PA: Muhlenberg College. 39-45.

Petrie, H.G. (1987b). Teacher education, the liberal arts, and extended preparation programs. *Educational Policy, 1*(1). 29-42.

—Reprinted by The Nelson A. Rockefeller Institute of   Government, State University of New York, No. 26 (Spring, 1987).

—Reprinted in Weis, L., Altbach, P.G., Kelly, G.P., Petrie, H.G., and Slaughter, S. (eds.), (1989a). *Crisis in teaching: perspectives on current reforms*. Albany, New York: SUNY Press.

Petrie, H.G. (1990). Reflections on the second wave of reform: restructuring the teaching profession. In Jacobson, S., and Conway, J. (eds.), *Educational leadership in an age of reform*. New York: Longman. 14-29.

---

3   See Petrie, H.G. and Oshlag, R. (1993), next page.
4   Revised and expanded (2011) Menlo Park, CA: Living Control Systems Publishing

Petrie, H.G. (1992a). Interdisciplinary education: are we faced with insurmountable opportunities? In Grant, G. (ed.) *Review of Research in Education, 18.* Washington, DC: American Educational Research Association. 299-333.

Petrie, H.G. (1992b). Knowledge, practice, and judgment, *Educational Foundations,* 6(1). 35-48

Petrie, H.G. and Oshlag, R. (1993). Metaphor and learning. In Ortony, A. (ed.), *Metaphor and thought,* Second Edition, (New York, Cambridge University Press) 1993, 579-609.

Petrie, H.G. (1995a) A new paradigm for practical research, in Petrie, H.G. (ed.), *Professionalization, partnership, and power: building professional development schools.* Albany, NY: SUNY Press. 285-302.

Petrie, H.G. (ed.) (1995a). *Professionalization, partnership, and power: building professional development schools.* Albany, NY: SUNY Press.

Petrie, H.G. (1995b). Purpose, context, and synthesis: can we avoid relativism? In D. Fournier (ed.), *Reasoning in evaluation: inferential links and leaps.* San Francisco: Jossey-Bass. 81-91.

Petrie, H.G. (1998). From 'my work' to 'our work', in Jacobson, S.L., Emihovich, C., Helfrich, J., Petrie, H.G., and Stevenson, R.B. *Transforming schools and schools of education: a new vision for preparing educators.* Thousand Oaks, CA: Corwin Press, Inc. 23-45.

Popper, K. (1965). *Conjectures and refutations: the growth of scientific knowledge.* 2nd edition. New York: Basic Books.

Powers, W.T. (1973). *Behavior: the control of perception.* Chicago: Aldine.
—Second edition (2005), revised and expanded, Bloomfield, NJ: Benchmark Publications

Powers, W.T. (1998). *Making sense of behavior: the meaning of control.* Bloomfield, NJ: Benchmark Publications.

Scheffler, I. (1960). *The language of education.* Springfield, IL: Charles C. Thomas.

Scheffler, I. (1965). *Conditions of knowledge.* Chicago: Scott, Foresman.

Toulmin, S. (1963). *Foresight and understanding.* New York: Harper and Row.

Weis, L., Altbach, P.G., Kelly, G.P., Petrie, H.G., and Slaughter, S. (eds.), (1989a). *Crisis in teaching: perspectives on current reforms.* Albany, New York: SUNY Press.

Weis, L., Farrar, E., Petrie, H. (eds.) (1989b). *Dropouts from school: issues, dilemmas and solutions.* Albany, New York: SUNY Press.

Weis, L., Altbach, P.G., Kelly, G.P., and Petrie, H.G. (eds.) (1991). *Critical perspectives on early childhood education.* Albany, NY: SUNY Press.

= Reprinted in this anthology

[1968]
Why Has Learning Theory Failed To Teach Us How To Learn?

The purpose of this paper is to suggest a novel answer to a tired old question. The question is: Why, despite the almost universally held belief that psychology and especially learning theory are the foundation sciences of education, have these "foundations" given such minimal support and assistance to actual day-to-day educational practice? And the answer which I will suggest is that, paradoxical as it may sound, learning theorists in psychology and practical educators are, for the most part, talking about two entirely different things.

I think it is abundantly evident that psychology, with the possible exception of psychometrics, has contributed little, if anything, to education. At any rate it is clear that learning theory, at once hailed as the best developed of the fields of psychology and at the same time the one field from which the most could reasonably be expected for educational purposes, has contributed next to nothing. For even Ernest Hilgard, one of the most respected learning theorists, and one who is interested in the problems of relating basic research in psychology to educational practice, clearly recognizes the paucity of contribution that learning theory has made. In both the 1964 NSSE yearbook,[1] of which he is the editor, and in the third edition of his own widely read book on learning theory,[2] Hilgard concludes with an apologetic for the seeming irrelevance of learning theory to education.

It will be instructive to see the kind of reasons Hilgard advances for this lack of relation, in order better to compare them with the answer I am proposing. His reasons for the lack of relation are essentially two. On the one hand is the general problem of development and application of theory which is common to all applied disciplines. On the other hand Hilgard believes that educators have generally not adequately specified

---

© Hugh G. Petrie. First published in Proceedings of the Philosophy of Education Society, 1968. 163-170. Reprinted in Broudy, H. S., Ennis, R. H., and Krimerman, L.I. (eds.), (1973). Philosophy of educational research. Chapter 2: Nature, Scope, and Strategy of Educational Research. pp 131-138. New York: John Wiley and Sons.

the tasks and the criteria of success for these tasks for basic theory to be of much use. And, of course, Hilgard's two answers are commonly accepted by psychologists and educators alike.

Without denying the importance of what Hilgard says, what I wish to do is to point out that problems of development and application and task analysis logically presuppose that the facts of learning are the same for the different learning theorists and for the educator. As Hilgard says, "all the theorists accept all of the facts."[3] That such a presupposition is indeed present is easy to see. We could scarcely begin to concern ourselves with development and application of theoretical results to concrete situations unless the facts of the concrete situations are of the same nature as the facts of laboratory-based theory. Nor would a more precise specification of tasks help in applying theory to practice unless the object domain of the task is the same as that of the theory.

For that matter, the supposition that all the theorists accept all the facts is not a surprising one. It is a fairly common piece of scientific folklore and just a simple restatement of the generally accepted principle that we can always draw a sharp and clear distinction between an observation language which reports the facts of our environment and a theoretical language which interprets those facts. Thus the presupposition is that there is a neutral data language upon which all agree and there are differing theoretical languages to interpret the data and over which there can be disagreement.[4]

And yet, there has recently arisen a serious challenge to such "obvious" presuppositions. It can be found in the writings of such men as N.R. Hanson,[5] W.V.O. Quine,[6] Stephen Toulmin,[7] and, perhaps best known of all, T.S. Kuhn.[8] These men have begun to argue that scientific theories are radically underdetermined by experience, and that although scientific theories must have empirical content—be testable by experience—they do not and cannot arise solely out of experience. It has been argued that what even counts as experience is essentially theory-dependent. That is, two scientists may look at the "same" thing and, because of different theoretical perspectives, may literally not *see* the same object. What is relevant for one theory may be totally ignored by another theory and even be logically incapable of being observed.

It should be emphasized at this point what a truly radical conception this is. It might easily be supposed that all that is being claimed here is that any science in fact focuses on certain features of experience to describe and ignores others. For example, classical physics, it has often been said, owed much of its success to having concerned itself with just the right physical properties, position and momentum, ignoring such properties as color and taste. If this is the sort of thing being claimed, then why all the fuss?

But if we stop here we miss the point entirely. For the "focusing" conception of science indicated above logically presupposes a kind of neutral experiential base upon which one may focus, now here, now there. Correlatively, a neutral observation language is also presupposed within which one could in principle describe all the physical properties of situations and events, leaving to the scientific theory the choice of those features to be covered by the theory. The non-favored features are still "there"; they are simply not deemed relevant.

However, it is the position of the view under consideration that no such neutral observation language exists nor can experience be described independently of theory—a radical view indeed.

Psychologists are not unaware of the problems of being constrained in their observations by the use of certain favored approaches and methodologies. For example, Underwood[9] has noted the unimaginativeness of many verbal learning experiments which seem often to return to the basic techniques of paired-associate experiments. However, most psychologists tend to treat such problems of constraint as problems in the psychology of methodology, assuming that with proper care and imagination they can be overcome. Without in the least attempting to minimize the psychological part of this problem, I want to be as clear as possible in suggesting that there may well be a logical and conceptual problem as well. In other words, it may be the case that all the care and imagination in the world may be unable to help an experimenter see a certain result if such results are not countenanced by the theory he explicitly or implicitly espouses.

If such a theory-dependency thesis of observation is indeed true, then it can easily be seen, at least in outline, how this might give weight to my contention that the major reason learning theory has been of such little help to education is that learning theorists and educators are generally talking about two different things. For most learning theorists, given the general pervasiveness of at least a methodological behaviorism, will see more or less mechanical stimuli and responses; whereas, most educators, given the teleological concepts of ordinary language, see goals and actions as purposive. Such a conception immediately shows the extent to which Hilgard was correct in asserting that a better task analysis is often a good way of bridging the gap between theory and practice. For if the task description can be given an S-R twist it would be easier to make the application. On the other hand, if the general results of learning theory are cast in teleological form, the application would again be easier.

Let me then pursue the theory-dependency thesis a bit further. An extreme form of the thesis would present us with a most radical kind of Whorfianism. For if each of us sees only what the theory we have enables us to see, and it is furthermore granted that everyone's conceptual scheme differs at least slightly from everyone else's, and finally, that our conceptual schemes are, in some sense, our theories of the world, then no one ever sees precisely what anyone else sees, and a rigorous notion of intersubjective confirmation or justification of some one theory is logically out of the question. Such an extreme view often seems to be implied by some of the things Kuhn says.

Fortunately, I do not think that such an extreme view is correct. For one thing it faces all the difficulties which any radical skepticism faces along with some of its own which I shall briefly mention. First of all, if this kind of theory-dependency thesis is even intelligible at all, it will be intelligible *on its own grounds* only in terms of some theory which determines observational categories sufficient for us to see the intelligibility of the theory-dependency thesis. It seems obvious that such an all-embracing metatheory is nothing more nor less than philosophy and thus that philosophical argumentation is appropriate to the theory-dependency thesis. For if the thesis actually asserts that it itself is outside the realm of any justification, even a philosophical justification, then quite clearly we can have no justification for accepting it, and yet equally clearly the thesis is capable of being argued about.

Second, even if we grant the extreme Whorfian version of the theory as a metaphysical possibility, we could not on epistemological grounds ever assert or deny this possibility. For as Quine has so adequately pointed out,[10] there is no way of deciding on the basis of the empirical evidence between someone's looking at the world radically differently and a mistake in translation. To make sense of the differences in conceptualization we do find, we must assume a tremendously large core of common conceptualization as a background.

Having concluded this much, however, we are still left with a reasonably strong version of the thesis. And this version states that there may be logically incompatible observational categories which are, nevertheless, philosophically basic and hence incapable of being decided between on empirical grounds, although philosophical argumentation would be appropriate. There is also a weaker thesis which states that within a single philosophically basic observational category, it is possible to have differing empirical specifications of what falls under that category.

What I would now like to do is to illustrate both the strong and the weak theses with reference to some of the changes which have occurred in the definition of a stimulus as learning theorists have moved from conditioning theory to discrimination learning to conceptual behavior.

Historically, hard-line behaviorists began by taking the definition of a stimulus to be in terms of physical events of some sort or other impinging directly on the organism, e.g., light waves hitting photoreceptors, or auditory nerves being stimulated. And indeed such a definition works well for typical conditioning experiments where it is fairly easy to determine what change in the carefully controlled laboratory environment will count as a stimulus, and it is also fairly easy to generalize on the stimulus.

However, once one enters the field of discrimination learning, not only must the subject be conditioned to some stimulus, but also he must learn in some manner what is to *count* as a stimulus. This involves problems of attention, focusing, stimulus patterning, and stimulus generalization which do not seem to occur at all in classical conditioning experiments. Now is not the time to enter into a detailed discussion of the experimental results of discrimination learning. Nor will I discuss whether or not these results can be accommodated within classical conditioning theory by means of some sort of selection and retention of repeated total stimuli defined in physical terms.[11] It will be sufficient for my purposes to note that discrimination-learning results have prompted many psychologists to retreat from the kind of hardline definitional behaviorism exemplified, for example, by Hull to a methodological behaviorism. A "methodological behaviorism," as I shall use the term, allows the introduction of any number of "mentalistic" intermediaries, or representations, or cues, as long as the introduction of such cues can be shown to have genuine explanatory power within the theory and as long as there is some observational test of such cues, no matter how indirect. Even Skinner verbally subscribes only to a methodological behaviorism, although he combines this with a further belief that on his system very few, if any, such mentalistic cues need to be introduced.

When one moves to the area of concept formation, the problems become even more acute. In discrimination learning single stimuli need to be discriminated one from another, whereas in concept formation whole classes of stimuli need to be discriminated from other classes. To see the problems involved in attempting to carry over the definition of a stimulus in physical terms as specified in conditioning theory to the physical definition of the class of stimuli which call forth a given concept one need only reflect on the incredibly wide physical dissimilarities involved in all the physical objects falling under the concept of a chair. The possibility of remaining within the bounds of a physical definition of the stimuli seems remote indeed.

As a result, more and more psychologists have tended to introject into the organism larger and larger parts of the environment to which the organism is supposed to be responding in discrimination and concept learning. And this is, of course, to come closer to the position which many philosophers and gestalt psychologists have long urged; namely, that an organism responds to what it *believes* the environment to be and not to what the environment actually is.

And yet, as has been pointed out by Kendler,[12] the whole process of a change in the definition of a stimulus from conditioning to discrimination to concept formation can still be considered to fall under a theoretical stimulus- response associationism. Thus despite the change in definition of the stimulus (and usually corresponding changes in the definition of a response), we still have the notion that any behavioral event can be *analyzed* in terms of an environmental feature (stimulus), some components of total behavior (response), and the association between the two.

In the sense, then, in which human behavior is considered analyzable in an S-R kind of way, we have an illustration of the weak sense of the theory-dependency thesis. For it will be recalled that the weak version of this thesis claimed that there might be differences in empirical specification of a single philosophically basic observational category. Thus we have the philosophical category of an S-R analysis of human behavior, and differing empirical specifications of this observational category ranging from physical definitions to cues internal to the organism. If the basic philosophical category is indeed of the S-R variety, then the criteria for deciding on the empirical specification of this category in different situations are, broadly speaking, empirical in nature. That is, we must await the results of the psychologists' investigations to tell us which ones are correct.

Nevertheless, it is still easy to see how, even under the weak version of the theory-dependency thesis, it might be difficult to translate the results of learning theory into educational practice. For it seems obvious enough that the practicing educator observes the educational process largely in terms which define the stimulus as internal cues; whereas the most reliable, if limited, results in learning theory come from seeing stimuli in terms of physical events—two widely different conceptions.

But now what if the basic philosophical category of a stimulus-response analysis of human behavior is wholly rejected? That is, what happens if the notion of a human action is actually unanalyzable in such terms and is either itself a basic philosophical observational category or at least cannot be analyzed in the causal terms of the S-R conception. Charles Taylor[13] has recently argued the latter while Richard Taylor[14] has argued the former. That is, both have argued on philosophical grounds that human action is essentially teleological in character in so strong a sense that the S-R conception sketched above is wholly inapplicable. What we now have is an illustration of the strong version of the theory-dependency thesis. For the claim by the two Taylors is that no matter how stimuli are defined they cannot, logically cannot, be used as an observational category for human action. And the reason is that human action belongs to a philosophical category different from that embodied in an S-R conception. Note, too, that the criteria for deciding between an S-R conception and a broadly teleological conception of human action are philosophical in character and hence must be decided on philosophical grounds.

Without deciding if ordinary language analyses actually yield the metaphysical results claimed for them, one can grant that the analyses of our ordinary use of action terminology are indeed teleological as claimed by the two Taylors. But if this is granted, and if it is further granted that practicing educators largely make use of ordinary language in describing the educational process, then it will follow that the theory embodied in ordinary language renders it logically impossible to observe human action in the educational process in the categories in which learning theorists state their results. And hence it is logically impossible, as long as ordinary terms are used as the basic philosophical category for the observation of human action, that learning theory as presently constituted could be of any relevance to education. For the basic philosophical categories of the two ways of looking at the world are incompatible and it will require a philosophical argument to settle the issue between them.

In conclusion let me make a few comments on this analysis. First, the framework I have offered gives *prima facie* promise of providing an explanation of how it is that learning theory has contributed what it has. Under my view one ought to be able to predict that principles of conditioning theory are most applicable in areas where our ordinary language concepts are not teleological, and least successful where such ordinary concepts are teleological, and indeed a glance at Hilgard's summary of just these items reveals a *prima facie* confirmation.[15]

Second, my own opinion is that the two Taylors are wrong in asserting that the teleological character of human action is such as to render it inexplicable in an extended S-R framework. However, this is essentially the philosophical controversy over whether reasons or intentions or motives can be causes, and it cannot be entered into now. However, as I have urged, the solution to this question must necessarily be a philosophical one.

Third, given an extended S-R framework, the isomorphism which has been noted by Suppes and Atkinson[16] between the recent mathematical S-R learning theories and certain cognitive theories is easily understood. The formal isomorphism could be proved because both fell within the broad formal framework of an S-R conception of human action although they may have differed in empirical specification of stimulus and response. A cognitive theory falling under a different basic philosophical conception could probably not be proved isomorphic.

Fourth, I have not argued directly for the theory-dependency thesis, but rather have assumed it to be in broad outline correct. It has seemed to me that such a view has been ably argued by others and has not been conclusively refuted. Thus, I believe it deserves to have some of its implications traced out in detail, and I consider the framework it provides for understanding the problems I have sketched in this paper to be a kind of indirect argument for the theory-dependency thesis.

Finally, despite the sweeping topics I have considered and the sketchy treatment I have offered of them, I believe I have made it at least plausible that there may be philosophical reasons for the seeming irrelevance of learning theory to education. I hope I have also been able to indicate the vast amount of work which remains to be done by philosophers of psychology and philosophers of education in this area.[17]

Notes

1. E. R. Hilgard (ed.), *Theories of Learning and Instruction,* Yearbook LXIII (Chicago: National Society for The Study of Education, 1964)
2. E. R. Hilgard and G. H. Bower, *Theories of Learning,* 3rd ed. (New York: Appleton-Century-Crofts, 1966).
3. *Ibid.,* p. 9.
4. *Ibid.,* p. 9.
5. N. R. Hanson, *Patterns of Discovery* (London: Cambridge University Press, 1958).
6. W. V. O. Quine, *Word and Object* (New York: John Wiley and Sons, 1960).
7. Stephen Toulmin, *Foresight and Understanding* (New York: Harper Torch Books, 1961).
8. T. S. Kuhn, *The Structure of Scientific Revolutions* (Chicago: University of Chicago Press, 1962).
9. B. J. Underwood, "The Representativeness of Rote Verbal Learning" in A. W. Melton (ed.), *Categories of Human Learning* (New York: Academic Press, 1964).
10. W. V. O. Quine, *op. cit.*
11. However, see Charles Taylor, *The Explanation of Behavior* (London: Routledge and Kegan Paul, 1964), for a sustained attack on the possibility that a simple extension of classical conditioning principles can account for the results of discrimination learning.
12. H. H. Kendler, "Concept of the Concept," in A. W. Melton (ed.), *Categories of Human Learning* (New York: Academic Press, 1964).
13. Charles Taylor, *op. cit.*
14. Richard Taylor, *Action and Purpose* (Englewood Cliffs, N.J.: Prentice-Hall, 1966).
15. E. R. Hilgard, *op. cit.,* p. 562-64.
16. P. Suppes and R. C. Atkinson, *Markov Learning Models for Multiperson Interactions* (Stanford: Stanford University Press, 1960).
17. The foregoing work has been supported in part by the Office of Education, Contract No. 0-8-080023-3669 (010).

[1972]
Theories are Tested by Observing the Facts —Or Are They?

I

"The first rule that we must be prepared to accept as we judge the relative merits of different [learning] theories is this: *All the theorists accept all of the facts."*[1] This view, enunciated by one of the most respected learning theorists, is, I believe, a view which is held almost without question by most psychologists, educators, and laymen. It is also, I believe, a view which is a most profoundly mistaken one. It is the purpose of this essay to indicate briefly why this view is mistaken and to show with a discussion of several examples what a pernicious influence this view has had on the development of psychological learning theory and most especially on the application of psychological learning theory to educational concerns.

The supposition that all the theorists accept all the facts is not a very surprising one. It is a fairly common piece of current scientific folklore and just a simple restatement of the generally accepted positivist principle that we can always draw a sharp and clear distinction between an observation language which reports the facts of our environment and a theoretical language which interprets those facts. Thus the presupposition is that there is a neutral data language upon which all agree and differing theoretical languages interpreting the data over which there can be disagreement.

And yet, there has recently arisen a serious challenge offered by such men as Hanson,[2] Quine,[3] Toulmin,[4] Kuhn,[5] and Petrie[6] to such "obvious" presuppositions. These men have begun to argue that scientific theories are radically underdetermined by experience and that although scientific theories must have empirical content—be testable by experience—they do not and cannot arise solely out of experience.

Originally published in: Philosophical Redirection of Educational Research, National Society for the Study of Education yearbook, 71st Yearbook. Chicago: University of Chicago Press. 47-73.

It has been argued that what even counts as experience is essentially theory dependent. That is, two scientists may look at the "same" thing and, because of different theoretical perspectives, may literally not *see* the same object. What is relevant for one theory may be logically and methodologically incapable of being observed under the presuppositions of a different theory.

It should be emphasized at this point what a truly radical conception this is. It might be supposed what is being claimed here is that any science in fact focuses on certain features of experience to describe and ignores others. For example, classical physics, it has often been said, owes much of its success to having concerned itself with just the right physical properties, position and momentum, ignoring color, taste, and so on. If this is the sort of thing being claimed, then why all the fuss?

But this would be to miss the point entirely. For such a "focusing" conception of science logically presupposes a kind of neutral experiential base upon which one may focus, now here, now there. Correlatively, a neutral observation language is also presupposed within which one could in principle describe all the physical properties of situations and events, leaving it to the scientific theory to pick out those features which are to be covered by the theory. The nonfavored features are still "there"; they are simply not deemed relevant.

However, it is the position of the view under consideration that no such neutral observation language exists nor can experience be described independently of theory. Thus, under this radical view the very categories of things which comprise the "facts" are theory dependent—the exact opposite of Hilgard's optimistic claim.

It should be pointed out that the discussion thus far has proceeded and will continue to proceed on the level of the *description* of experience. That is, I have taken what has been called the linguistic turn.[7] Now it has been argued that such a move is ultimately unacceptable.[8] According to this line of criticism I should not even speak of our *experience* of the world, already a transcendental turn, but rather of the world itself if, indeed, I wish to say something about the world. I cannot here enter into this very interesting discussion. My reason for avoiding it is that when Hilgard says that all the theorists accept all the facts, he clearly means that psychologists do not argue over the experimental data which is reported in the journals. Clearly this is on the level of the description of the psychologists' experience. Thus I will also not here worry over the exact ontological status of the troublesome concept of a fact.[9] Rather,

I shall content myself with treating as a fact those linguistic entities which appear as "data" categories in the typical experimental psychology literature. These are clearly the facts to which Hilgard is referring and they are at least closely related to the "facts" that philosophers discuss.

Thus when I speak of "observational categories" I have in mind such things as "trial," "bar press," and "series of nonsense-syllable responses." These are the kinds of terms which appear in journal articles and which count as "observable" in some sense or other. On the other hand, when I speak of "what we see," I mean the experiences by virtue of which we go on to apply the foregoing observational categories.

II

An extreme form of the theory-dependency thesis of observation would present us with a most radical kind of Whorfianism. For if each of us sees only what the theory we have enables us to see, and it is furthermore granted that everyone's conceptual scheme differs at least slightly from everyone else's and, finally, that our conceptual schemes are, in some sense, our theories of the world, then no one ever sees precisely what anyone else sees, and a rigorous notion of intersubjective confirmation or justification of some one theory as over against another is logically out of the question. Such an extreme view often seems to be implied by some of the things Kuhn says.[10]

I do not think that such an extreme view is correct. For one thing, it has all the difficulties which any radical skepticism faces, in addition to some of its own. First of all, if this kind of theory-dependency thesis is even intelligible at all, it will be intelligible *on its own grounds* only in terms of some theory which determines observational categories sufficient for us to see the intelligibility of the theory-dependency thesis. It seems obvious that such an all-embracing metatheory is nothing more nor less than philosophy and thus that philosophical argumentation is appropriate to the theory-dependency thesis. For if the thesis actually asserts that it itself is outside the realm of any justification, even a philosophical justification, then quite clearly we can have no justification for accepting it, and yet equally clearly the thesis is capable of being argued about.

Second, even if we were to grant the extreme Whorfian version of the theory as a metaphysical possibility, we could not on epistemological grounds ever assert or deny this possibility. For as Quine has so adequately pointed out,[11] there is no way of deciding on the basis of the empirical evidence between someone's looking at the world in a radically different way and a mistake in translation. To make sense of the differences in conceptualization we do find, we must assume a tremendously large core of common conceptualization as a background.

But if such an extreme construct of the theory-dependency thesis is untenable, nevertheless there remains a version of the thesis which is of vital importance to the relationship between epistemology and learning theories. For the traditional ground of objectivity for all empirical theories and learning theories in particular—viz., appeal to observation—is itself asserted to be theory-bound. But this means that the role of observation in providing all or even part of the objective basis for theory stands in need of radical reanalysis. One cannot *simply* appeal to observation to settle empirical disputes. A theory might perhaps be refuted "objectively" on rationalistic grounds. The point is that such objectivity would be inconceivable to those who, like Hilgard, look upon observation as *the* objective ground of theory.

On the other hand, the rejection of the extreme subjectivist interpretation of the theory-dependency thesis places me somewhat in line with at least the thrust of Hilgard's remarks. For if it is not true that any theory is as good as any other, then there must be *some* way or other of providing an objective ground for choosing between them even if this ground is not a neutral observation language. What is right, therefore, about Hilgard's remarks and what should be recoverable in an appropriate reanalysis is that theories are *commensurable* in some way or other.[12] In a way, it is the purpose of the rest of this essay to begin exhibiting in use varying modes in which we can objectively compare theories without appeal to a neutral observation language, although, of course, on investigation *some* sort of appeal to observation reanalyzed will be made. For certain purposes and within the bounds of certain presuppositions, it is true that observation grounds empirical knowledge. For this reason I will occasionally lapse into language reminiscent of Hilgard. I can only hope that such lapses will be clearly justified by my having made explicit my purposes and presuppositions.

I should also note here that even a mild form of the thesis of the theory dependency of observation has a paradoxical sound to many ears. For example, it might well be urged against some of the things I will say in the sequel that these remarks are themselves theory dependent and hence not compelling for anyone who does not accept my theories. In reply, let me repeat that I think such an objection has weight only against the extreme form of the theory-dependency thesis. And I have already rejected that form of the thesis. In addition I have, like Quine, rejected the possibility of a philosophical justification of epistemology.[13] And, like Quine, I will make free use of science in treating problems of "epistemology," or "critical methodology," or whatever you will. Rational discussion can still take place against the background of the presumptions of philosophy and science. It is only if it is thought that my remarks are an attempt to justify this background a priori in a way which is independent of the theory-dependency thesis that misunderstanding might arise. I make no such sweeping claims. The scientific and other results that I use *are* subject to the theory-dependency thesis. But, since I have not ruled out the possibility of some of these results being "better" than others, my position results in no contradiction.

But now what reasons are there for believing even the modified version of the theory-dependency thesis to the effect that objectivity is not solely guaranteed by neutral observation. The reasons are complex and varied and can be found discussed at great length in some of the authors already cited, e.g., Kuhn, Hanson, Quine, Petrie, and Toulmin.[14] For my purpose here I wish briefly to run over two lines of argument. On the one hand, there is the apparent failure of all programs to identify the basic particulate materials of observation. In philosophy, this failure is exemplified by the downfall of sense-data theories and phenomenalism as viable answers to the problems of perception.[15] In psychology, the work of gestalt psychologists in perception bears ample testimony to the importance of individual conditions on what is perceived.[16] Various experiments on "set" also indicate the extreme difficulty if not impossibility of finding a unique "neutral" or "objective" description of observation. The very categories of what can be perceived seem theory dependent.[17]

On the other hand, if there are no basic perceptual categories to which we can refer observation or into which observation could be analyzed, then it would seem that the criteria for membership in a purportedly "basic" perceptual category must be more or less indefinite and amenable to modification. One could avoid this looseness only by giving up on classification and reverting to simply proper naming. But once that stage is reached, the very idea has been abandoned of finding some public, repeatable, intersubjectively verifiable observation that could serve as the objective basis for theoretical interpretation. As in sense-data theories, complete specificity is purchaseable only at the price of being unable to say anything *about* the specified item.

III

These two lines of argument come together in a most illuminating way in a consideration of operational definitions. Operational definitions are crucially relevant to a discussion of the theory-dependency thesis of observation because it is just here that most theorists of Hilgard's persuasion take their stand on the distinction between the facts and what can be inferred on the basis of the facts. Operational definitions are the point at which theory makes contact with objective experience, and yet, as I shall urge, taking the theory-dependency-of-observation thesis seriously leads to a relativization of operational definitions which is in conflict with their purported role as objective anchors of theory in experience. Furthermore, this necessary relativization of operational definitions helps explain a number of current methodological problems in learning theories.[18]

Although there are many conceptions of what an operational definition should be and at least as many problems relating to it as there are conceptions, I shall here need only a very general notion of an operational definition. I shall consider an operational definition to be the association of a definite, observable, testing operation with the term being defined as its (the definiendum's) criterion of application. This formulation should suffice for my purposes.

The first thing to notice is that in order to establish the reliability of the operational definition, one must be able to identify both occasions of the application of the testing operation and what happens in each instance as a result. For example, one must recognize any number of occasions of the taking of an intelligence test if one is using such an operational definition of intelligence. Although these events all fall under the concept of taking the test, they also differ among themselves in an indefinite number of other characteristics or else they would not be separable events. This is the exemplification of the requirement, just discussed, that the events be classified into some category or other and not just named. What this also shows is that there are two distinct logical roles which must be played in an operational definition. These are: first, the role of observable terms, namely, the test operation and its result, and, second, the role of theoretical terms defined by the operational definition. The observable terms must be assumed to be reliably identifiable from occasion to occasion. For it is only if these terms are identifiable that we can go on to ask the further empirical question of whether or not these two observables go together. We must assume that there are some persisting, underlying properties of one form or another by virtue of which we can classify two or more events as being events of the kind determined by the properties. Thus the problem of operational definitions is not that of the distinction between observables and underlying properties. *All* operational definitions assume some sort of underlying properties as the basis of being able to apply and reapply the terms functioning as observables. The question is, rather, as to the particular choice of terms to utilize as observables.

But now the other line of argument mentioned above concerning our inability to find a unique set of observables comes into play. I have argued that the observation terms in an operational definition serve the logical role of picking out those categories whose criteria of application we assume give us no trouble. But if we cannot locate or describe any absolute or independent set of these categories which could serve as ultimately observable for all operational definitions, then we are forced to admit a relativity of operational definitions. What one theorist treats as an observation term for his operational definitions, another might believe needs itself to be operationally defined, in which case it would play the role of *a* theoretical term. One man's reliability problem may be another's validity problem.[19] In short, the theory dependency of observation infects operational definitions themselves, with no obvious way to settle on some single set of observation terms. It is indeed a nice philosophical problem as to just how, if at all, one *can* justifiably settle on a set of observation terms for any given investigation. I have argued elsewhere that at least a part of the answer is that we must accept the well-confirmed empirical laws of the moment even though from a logical point of view they may be systematically misleading. That is, we accept as observation terms only those concepts which we have no serious theoretical or empirical reason to believe unreliable. However, such concepts may well prove not to be descriptive of the world. For example, people spoke (and we still do) of seeing the sun rising in the east, even though the horizon is turning away from us.[20]

The application to actual practice of this epistemological doctrine of the relativity of observation to theory is so simple and so widespread as to be almost unnoticed because of its ubiquity. The use of an ordinary control group (plus, of course, extraordinary control groups) is an example. Let us clear our minds of all the bias and prejudice and methodology we have learned and reconsider the situation. A psychologist wants to find out if a certain treatment has a certain effect. He tests, administers the treatment, and tests again to find out if the treatment had any effect. (One could really get down to bare bones by omitting the pretest, which is itself a kind of control group, but the obvious possibility of already existing capacity prior to the treatment makes this supposition strain credulity a bit too far.) On this model we see the pretest as simply a measure of current competence. However, if our "theory" (it need not be terribly precise) enables us to see the pretest also as an opportunity for practice, then it might well occur to us that the resulting effects could be as easily due to the practice as to the treatment. Hence, with a creative methodological leap someone might, looking at the situation in this way, design the experiment with a control group to whom the pretest and posttest are administered but for whom the treatment is omitted. For the experimental group, treatment is included and the results of the two groups compared, and if a difference is noted, it is ascribed to the treatment.

It would be well to point out here that, commonsensically, there is nothing more "observable" about a pretest than practice. They are just different descriptions of the same event. Some, under the effects of the long-standing positivist mythology, might object that the pretest must be inferred to be practice as a result of the control group experiment. Unless it actually has an effect on the result it cannot be counted as practice and hence is not observable *as* practice. There are two responses to this objection. First, it is a piece of ad hoc linguistic legislation, for it follows from this view that one could never discover that practice did *not* have an effect on later performance since practice is defined as precisely that which does have such an effect. Of even more importance, however, is that this just pushes one step back the ultimate decision as to what shall play the role of observation and what the role of inference. After all, it might equally be argued that the category of pretest must itself be inferred. For it will be a pretest only if we can reliably establish that the subjects are actually displaying their competence and not playing a game with the experimenter which consists of giving as many wrong answers as they can.

Nor will it do to treat the above discussion as a reason for adopting a most extreme form of behaviorism. For even if we eliminate such mentalistic notions as "playing a game," the behaviorist too must establish the reliability of his operational definitions in terms of, for example, gross muscular movements. Even here, mistake is possible and confounded results can occur. The point is simple. At *any* level a question can be raised about the reliability of the application of an observational term, and in the absence of a unique specifiable observation language, the terms which *are* treated as observational will depend on the implicit or explicit theory.[21]

The problem with the foregoing example is that the theory (or common sense) which determines the use of control groups is by now so well confirmed and widely accepted that it may be more than a little difficult for people to be convinced that someone might want to fiddle with its observation terms. For this reason let me give another example. Even with the use of a control group of one form or another, the possibility remains that the attribution of the result to the treatment, though dictated by the logic of the experimental methodology, might yet be in error. One of the most striking recent examples of this is the work of Rosenthal.[22] Roughly the situation is this: Rosenthal has demonstrated the widespread effect of experimenter expectation and the effect of the affective tone of the experimenter on experimental results which had up until that time been wholly ascribed to the experimental treatment. For example, in my terminology the observation term "administration of treatment" was not at all reliably connected with the results. A more appropriate observation term was, e.g., "administration of treatment with certain affective tone." We do not merely see a certain event, we see it under certain aspects.[23] And once again, although the *role* of observation terms must be played by some of the concepts, it is a theoretical-empirical question as to which concepts can best play this role. No unique set of observation terms seems to have any more a priori justification than any other set.

IV

What I now want to do is to illustrate the thesis of the theory dependency of observation in five selected problem areas—the question of what is learned, the problems of hierarchical structures in learning, the problem of latent learning, the law of effect, and the area of developmental psychology. In each of these areas I will attempt to show how various theories and approaches end up being very nearly incommensurable because of differences in what they allow to be seen. I shall not attempt any detailed exposition of various learning theories because very few are coherent or complete enough to warrant exposition, and because I suspect almost no one would be willing to be identified with such an exposition. I am more concerned with illustrating the theory dependency of observation than I am in evaluating alternative learning theories, and I believe this can most easily be accomplished by concentrating on several issues in detail.

The controversy in psychology over the question "what is learned?" is sometimes presented as a dispute between those who would assert that some sort of central underlying process or disposition is acquired in learning as opposed to those who believe that various pieces of observable behavior are acquired. This is, however, misleading. For at least some of the controversy involves the choice of an appropriate set of observational categories without necessarily involving "underlying" processes or "hypothetical constructs" or "intervening variables" at all. It was pointed out by Campbell that the acceptance of the position that all we have to go on in constructing our psychological theories are the responses of the organism does not thereby commit one to supposing that no central states can be legitimately inferred.[24] Nor does this imply that there is only one way of observationally categorizing the behavior which must serve as our grounding. In short, what is learned may go considerably beyond any simple categorization or combination of observed behaviors.

Consider one of the experiments discussed by Campbell. In this experiment a conditioned finger movement was obtained through pairing a shock and a tone. The shock could be removed by an extensor movement of the finger. What was learned? An extensor finger movement or withdrawal of the finger? The question is open to experimental investigation. Turn the hand over and repeat the experiment. Now an extensor movement does not remove the shock, but finger withdrawal will. Ninety percent of the subjects withdrew their fingers, 10 percent continued the extensor movement.

It would seem that what was learned by most of the subjects was an action as contrasted with mere behavior. But now many behaviorisms such as Thorndike's connectionism, Guthrie's contiguous conditioning, along with the current eclectic functionalism, would almost all be committed to an explanation in terms of first having learned an extensor movement, next having transfer of training of a muscle group, and a response gradient reaching the new muscle group. Doubtless something like spread of effect would also be invoked. On the other hand, cognitive theories such as Tolman's sign-learning theory or some of the newer information-processing theories would see an action as the observational concept which can be applied to any number of quite dissimilar muscle movements or pieces of mere behavior.

For the associationists the units of behavior tend to be particulate bits of mere behavior and complex behavior must be inferred. And yet in this case their methodological bias seems to require a whole host of strange-sounding entities such as gradients, transfers, and so forth. And even if these "entities" do not have ontological status, nevertheless they stand for very complex processes which seem more and more ad hoc or else seem to require a retreat from a molar behaviorism to physics and physiology in order to be completely clear and unambiguous. The molar criteria for application of the observational categories in the case of the finger withdrawal become tremendously complex, and the predictions which can be made are extremely limited in scope.

The cognitivists, on the other hand, use a set of observational categories in the finger withdrawal case that have as criteria of application some resulting state of affairs which might well be reached in a number of different ways. It would seem far more appropriate with the strictures of a molar behavior theory simply to grant that what is learned is to withdraw one's finger at the tone and that this response is directly observable. The predictive power of this view would seem much better, at least in this case.

Skinner's operant analysis gives us yet a third view of the situation. Skinner's observational categories tend to include as a part of the criteria of application the particular reinforcement schedule applied to the operant behavior, i.e., to the behavior emitted naturally. Given this particular experiment, an operant analysis may not even be possible since the language is that of classical conditioning—usually a much coarser-grained language than is needed for operant conditioning. We simply do not know what behavior was shaped by the particular reinforcement histories. What is learned for Skinner is a sequence of behavior precisely dovetailing into the particular reinforcement schedule which was operative.

Can anything be said about these three different views of what is learned? First of all, no one of them can be said to be "closer to the facts" than any other, for if the preceding discussion is at all correct, the *facts* are different for each of these views at least in the sense of what plays the role of observation and what is inferred. (This is not to say that we cannot make some translations. My descriptions of the situation hopefully provided clues for such translations. It is because I earlier rejected the extreme form of the theory-dependency-of-observation thesis that I can now make sense of some kind of similarity among the positions.) Nor will it do to say that one position makes use of underlying properties while another does not. I have already argued that all are committed to underlying properties which allow for reidentification, albeit the properties differ from theory to theory. According to my previous discussion, the only plausible way of deciding between the theories is by undertaking a thorough theoretical cum empirical examination of their presuppositions, simplicity, predictive power, and so on.

It might be useful to point out here that on the above criterion for evaluation of the various theories, common sense as reflected in ordinary language (the theory of common sense) probably stands head and shoulders above all the other theories. Of the three, cognitive approaches seem best for this particular case insofar as they share, with common sense, observation categories which are of actions rather than of mere movements. Categories applied in terms of intentions and results are often much more reliable than those applied in terms of reinforcement schedules or spatiotemporal bodily movements.[25]

Let me take another example. Consider two baseball pitchers, one left-handed, one right-handed. Suppose further that one has learned the game wholly by playing, reinforced only by natural contingencies; whereas the other has had the benefit of sustained coaching. Now suppose in a game both of them pick a runner off first base. It is hard for me to imagine two more different reinforcement histories and sets of muscular movements, and yet, in terms of action categories, these two men have quite obviously done the same observable thing, picked a runner off first.

It is somewhat surprising to realize that behaviorism, that general program in psychology most concerned about getting as close as possible to the observable facts of the world, uses observation categories which are, in so many cases, so very far from the simplest, most reliable categories they could use—viz., the action categories of ordinary language. The claim that human action is just as observable as human movement is probably a defensible part of the emphasis put on *verstehen* by Max Weber.[26] To observe human behavior with *verstehen* is to see it as action and not mere movement from which action must be inferred.

But if the foregoing constitutes a kind of defense for the use of observational categories of human action in learning theory, at the same time it seems to ignore the 10 percent of the subjects in the finger experiment who continued to utilize extensor movement. *They* did not learn to withdraw their finger. The point to be noted here is that various kinds of conditioning theories do seem to account more adequately for certain kinds of learning than do cognitive theories. We sometimes do learn things "by rote."

The question then arises as to whether or not one needs to invoke a hierarchy of types of behavioral units, or whether these apparently different types of things can be explained in terms of one another.

Consider first the attempt to explain away the more cognitive type of behavior in terms of mere muscle movement. Taking a theory such as Hull's, we find that in order to account for what the cognitivist would call insightful or rule-governed behavior, use is made of such concepts as stimulus generalization, fractional antedating goal response, and habit-family hierarchy. The first concerns essentially how any number of physically quite dissimilar stimuli can all be said to be associated with a certain class of responses. For example, think of all the physically dissimilar instructions that might be written, printed, spoken, etc., to get someone to respond. Under a Hullian type of explanation one should be able to account for this by means of a generalization from the originally learned stimulus. The fractional antedating goal response is meant to be used with the habit-family hierarchy to explain why any number of pieces of behavior can be utilized to attain a specific goal. (Compare my example of the baseball pitchers.) The idea is that because of various reinforcement gradients different specific behavior patterns are differentially conditioned to the goal stimulus. Thus under certain conditions one pattern will emerge, while under other conditions another will. This integrated set of patterns constitutes the habit-family hierarchy which is derived from the elementary behavioral units.

Schematically the problem with this kind of handling of the different types of things which are learned is that of specifying the range of the stimulus generalization and the extent of the integration of the habit-family hierarchy. What tends to happen is that either the principle of generalization of the stimulus begs the question by covertly importing the very cognitive term to be defined, or else remains totally unconvincing empirically. In fact, it is only the continual equivocation between sneaking in the terms to be defined and moving back to empirical accounts that has allowed such a notion to survive as long as it has.[27]

On the other hand, however, associative learning is not likely to be assimilated to cognitive learning. The lower animals simply do not seem to display the requisite cognitive functions and yet they do learn in apparently associative ways. There is also a philosophical reason for supposing that not all learning, even in humans, makes use of concepts like the gestaltist's "insight" and "organization." Wittgenstein points this out in discussing why reference to obeying a rule is not ultimately explanatory.[28]

Briefly the reason is that rules can always be variously interpreted and until we ground the infinite regress of rule, interpretation of the rule, another rule to enable us to understand the interpretation, yet another interpretation, and so on, in some kind of actual behavior in the world (form of life), we will be unable to explain the correct interpretation of the rule.[29]

What emerges is the necessity for a nonreducible hierarchy with genuinely different criteria of application of the terms at the various levels. What the associationist needs to recognize is that no matter what the fine details of the associative bonds formed between bits of behavior, or no matter what the reinforcement history, the criteria of application of the concept of what is learned (and what will thus be reinforced) do not, in some cases, refer to these bonds or this history. The concept of what is learned is often analyzed in other ways, such as in terms of intentions, or actions, or results of the behavior, independently of its fine details. But such concepts can themselves become the units of behavior, i.e., the observables, for a new round of learning. Concomitantly, the gestaltist or cognitivist must realize that to characterize all learned behavior with such observation terms seems to ignore the 10 percent who learned the extensor finger movement. But even more importantly, it gives up the explanatory value of classifying some learning as insightful. For such a classification is explanatory only as long as it is used to mark off a different set of observation categories from the ones traditionally used by the associationists. A part of correct explanation is good classification as long as the classification marks a real difference.

It is no accident that information-processing theories of learning sound very "cognitive" in their language. This is at least partly because these theories have incorporated the notion of a hierarchy from the beginning. They take seriously the fact that the higher-order processes not only can represent lower-order processes in one way or another, but also that these higher-order processes have other features of their own. Just as a mathematical theory may represent reality, there are syntactic and purely mathematical features of the theory which are incidental to its representing function but nevertheless quite real. Thus the categories in the higher levels may represent the lower levels, but not be identifiable by means of reference to the lower levels. The hierarchy is nonreducible. On the other hand, however, information-processing theories are just as "mechanical," "hardheaded," and "nonmysterious" as the most dedicated behaviorist could wish. The continuing success and fruitfulness of computer simulation techniques utilizing such theories have graphically shown that low-order mechanical reactions can be so organized as to produce highly organized "meaningful" behavior. In short, simulation has shown the associationist he need not fear mentalistic vocabularies and has shown the cognitivist that his higher-order concepts can often be realized in a totally nonmysterious way.

The point is that if we attempt to describe all experience in associationist categories, certain features, e.g., the higher-order functions, seem to get left out. Associationist theory simply seems to have no room for a hierarchy even if the hierarchy is ontologically unobjectionable, as in information-processing theories. Cognitivist theories, however, seem to remain suspended in the air when an attempt is made to apply their observation categories to everything. Both associationists and cognitivists must be brought to realize the blinders which their theories impose on their observational categories.

I turn now to the phenomenon known as latent learning. This refers to the fact that in a variety of situations behavior goes on which does not at that time seem to result in any learning, i.e., no activity is originated or changed at the time, but is such that at some later time the learning is elicited. For example, rats able to explore a maze without reward begin, as soon as they are rewarded, performing as well as rats trained all along with a reward. A more homey example is the student studying for a test who does not perform until the test occurs.

This phenomenon is often marked by transformational linguists as the distinction between performance and competence. Although *some* kind of performance must usually be used to determine if competence has been gained, this need not always involve any straightforward exercise of the competence. For example, we might simply accept the performance of an actor who is trustworthy and who claims to have memorized his lines.

Although competence is probably not one of the observation categories of even the transformational linguist (i.e., it is inferred for him, too), it is much easier for him to handle since he does not operationally define competence in any single-track sort of way. Given that certain kinds of adaptive actions *are* observable, he simply postulates the competence as one of the hierarchical entities which help give rise to the actions. The competence has a life of its own independent of the actions which occasionally exemplify it. The justification for such status is the theoretical integration given to the actions supposed to flow from a certain competence.

A similar move is not nearly so easy for associationists to make. In the first place, they tend to define completely their inferred terms by means of operational definitions. This then cuts off any independent life for these terms. The inferred term *is* the operational definition and it makes no *sense* to ask if there are other operational definitions of the inferred term or whether the original definition should be modified. Such questions presuppose that the inferred term has some use beyond its operational definition. But even if the associationists allowed themselves a bit more freedom in the ontological status of their inferred entities, they would still face problems in accounting for latent learning because of the limitations of their observation categories. Let us see why.

Many associationists adopt a drive-reduction view of reinforcement. That is, crudely, they have observational reasons to believe in certain kinds of physiological and biological drives, e.g., hunger, thirst, and so on. Reinforcement occurs when one of these drives is reduced. Thus the drive causes behavior. When a certain pattern of behavior succeeds in reducing the drive, that pattern is reinforced and learning has taken place. Now the problem begins when the independently identifiable physiological and biological drives seem unconnected with the phenomenon of latent learning. In fact, that is just another way of specifying the area of latent learning. It is behavior continued and learned without the customary drive reduction. Typically, the associationist postulates new drives, as, for example, curiosity. The drive-reduction theory of reinforcement remains in effect and all is well. But not really. For given the behavioral observational categories of the associationists, the *only* indication of the new drives postulated to account for latent learning is the very behavior which manifests latent learning. There is no independent way to specify the drive. The assertion that there is a law-like connection between the drive and behavior which reduces the drive becomes tautological. With the limited observation categories of the associationist, there seems no plausible way to avoid such a result.

Notice here, however, that Skinner's notion of operant behavior seems to avoid the problem. For Skinner's basic observational category of operant behavior does not require him to see the behavior as caused. Where other associationists see two events, Skinner sees but one.[30] Thus Skinner is obligated to explain the change of behavior but not the origination of behavior. For *what* is observed is not just behavior, but behavior of a certain kind, i.e., of the kind connected with the particular reinforcement schedule.

The cognitivist also avoids the problem by not observing merely "action," but rather action of a certain kind. His criteria of identification refer to results and intentions rather than to reinforcement schedules, but in both cases a level higher up in the hierarchy is treated as containing the observation terms, thereby making inferences to higher levels a bit more plausible. In both cases, however, the directions which the theories can take in meeting their problems and the limitations they encounter seem closely bound up with what the theory says is observable.

The apparent circularity of the explanation of latent learning by the drive-reduction theory of reinforcement is similar to the oft-repeated charge that one of the major laws of learning, the so-called law of effect, is also circular. Although the law of effect is stated in innumerably different ways, examination of just one of these should suffice to show the possible pitfalls. Consider the following ". . . a learnable response followed by a reinforcing event (stimulus, state of affairs) will receive an increment in its strength or probability of occurrence." [31] The charge is then made that unless we can independently specify what a reinforcing event is, the law of effect becomes a definition of a reinforcing event and hence circular. This charge, however, generally does not pay sufficient attention to the subtlety with which associationists deploy the law of effect. In the first place, the law will be truly circular only in case there is not *any* independent description of the reinforcing event other than that it increases the strength of the learnable response. Of course, most of the time there is. What this exemplifies is the general principle that logical relations such as tautologicality depend on the descriptions used rather than on the referents of these descriptions.

Suppose, for example, that "*a* caused *b*" is a true singular causal statement—a paradigm case of an empirical statement. But, ex hypothesi, "*a* = the cause of *b*," and by substitution we get "the cause of *b* caused *b*," an analytic statement in which the description of the cause is *not* independent of the description of the effect. Yet one is under no compulsion to assert that the connection is not empirical.[32] The principle is generalizable.

Thus to characterize an event as a reinforcer is at least partially to claim that it is observably identifiable in terms of its effect on certain kinds of learnable responses. But that is *not* to say that the reinforcer may not be identifiable in some other way as well. Under this view it is possible that a reinforcing event may not reinforce. And there is no paradox here. The situation is similar to the lack of paradox in saying,

"Jim was shooting baskets yesterday, but he didn't shoot (make) a single one." In both cases the activity is identified partially in terms of what we know generally are the results of, or characterize, the activity. We are using observation terms which generally have a fairly wide range of application in the sense of classifying many otherwise dissimilar events as events of the same kind. The law of effect thus provides a learning-theoretic set of observation categories for behavioral events that can also be described in other terms which could in turn show the law of effect in particular cases to be empirically false.[33]

Such a view becomes even more plausible if we note that whether a reinforcer really reinforces depends on other things than the fact that it usually does so. It might also depend on the subject's noticing it, for example. The situation is analogous to the fact that a perceptible object will sometimes not be seen, even though it is in the subject's field of vision, unless other conditions are right, e.g., the subject is paying attention. Once more the influence of theory on observation terms is apparent.

All the preceding examples do exemplify, I believe, the relativity of observed and inferred and the theory dependency of observation. Nevertheless, it is probably the case that because many readers will themselves be committed to a particular theory, the phenomena I have taken to illustrate the theory-dependency-of-observation thesis will be seen by them as simply my pushing a particular substantive point of view (despite the fact that I have tried to accuse nearly everyone of parochial observation at one point or another). Part of the cure for this simply involves going over the material again and again and trying to take the other person's viewpoint. If you are an associationist, try to see (not infer) actions instead of movements in some instances. If you are a Skinnerian, try to see behavior through observational categories that do not have the particular history of reinforcement as part of the criteria of application. If you are a cognitivist, try to see that certain rudimentary behaviors need to be learned associatively to provide the basis for the application of your own (differing) observational categories. If you are a critic of associationism, try to see certain events *as* reinforcers rather than seeing one event, then another event, and inferring reinforcement.

In addition to such reiterations of what has gone before, let me offer one more example of the influence of the theory dependency of observation. This example is from developmental psychology, and because of the way it involves our most philosophically basic observational categories, it may be somewhat more persuasive than the preceding.[34]

If, following Kessen, we take cognitive psychology to be concerned with reality and man's representation of reality and if we attempt to explain how a child can grow up to master mathematics, physics, psychology, and even philosophy, it is clear that our analysis must necessarily be constrained by our own view of the world.

The problem in a nutshell is that we as adults define the terms and set the developmental goals, in terms of which we describe and evaluate a child's progress. It is the reality as adults structure it that we foist off on the child. There are some nice philosophical puzzles here about whether we can logically avoid such an imposition and to what extent we are justified in believing that our own adult conceptual scheme adequately represents reality, but although such questions are clearly relevant, they have often been discussed before. What I want to do here, however, is to assume that somehow, perhaps evolutionarily, the adult conceptual scheme is adaptive to, if not closely representative of, reality.[35]

With this assumption in hand I want to call attention to two current controversies that fit into the problem structure of developmental psychology. I refer to the revival of the nature-nurture controversy by the ethologists[36] and to the women's liberation movement.

Turning first to the ethologists, quite clearly one of the major problems is whether we are to see certain kinds of human behavior as essentially aggressive, or are we to see aggression as aberrant and caused by some lack of genetic determination or upbringing. Or is it some combination of the two? I do not wish for a moment to endorse what I find to be oversimplistic global explanations by the ethologists. What I do wish to do, however, is to point out that at least a part of what they are asking us to do is to *look* at the world in a different way—a way conditioned by their theories. The questions they raise are as much about the categories of observation as about theoretical interpretations. To give just one example—if aggressive behavior is "natural," then aggression can be counted as one of the basic physiological drives and used in explanations of other aggression-related behavior. If aggression is not natural, then *it* must be explained in some way by other drives or stimuli which are themselves assumed to be natural.[37]

The case of women's liberation is even more fascinating. Discounting the more rabid proponents, there is a most intriguing picture presented by the movement. Essentially, they are saying that the categories of femininity presented by the dominant male culture are *not* based on basic biological or physiological characteristics, but are rather the result of the cultural picture the adult male has of what a truly feminine human being should be like. The pervasive character of the picture is explained by the fact that women become "willing, unknowing participants" in their own subjugation. One of the interesting points here is that even after one attempts to determine empirically what attributes can be attributed to upbringing, the movement continues, at least in some instances, to make out a plausible case that the very categories in which the experiments are carried on are themselves male-culture–bound. For example, it is at least plausible to entertain the hypothesis that some of the physiological differences one finds between male and female and which show less strength on the part of the female are traceable to differences in the amount and kind of physical exercise our culture deems important for the two sexes.

Notice that I am not talking about the fairly obvious sociocultural differences in treatment of male and female, but rather about the basic categories of masculinity and femininity themselves. Do we start by seeing human beings and determining what differences there are, or do we start by seeing males and females and determining what similarities there are?

The analogy with child development is a good one. We cannot simply *use* without examination our basic categories of time, space, causality, etc., in describing the attempts of the child to construct these very categories. If we do, somehow we will likely miss the most important features of the process of construction; namely, are there reasons, other than the fact that adults reinforce their use, for developing the categories we do. The case with masculinity and femininity is similar. We cannot just use, without explicit awareness, attention, and justification, the present categories of masculinity and femininity to describe the attempts of women's liberation to question these very categories. Indeed, we must start somewhere, but the women's liberation movement has shifted what they will count as observable and what they believe must be inferred, and for us stubbornly and dogmatically to insist on a different set of observation categories even backed up by a simple elaboration of the theory which determines *our* categories is to miss the point entirely. Once more the relativity of observable versus inferred and the theory dependency of observational categories shows us why there is so much heat and so little light. The opponents simply talk past one another.

V

What methodological morals can be drawn from all this? For one thing, I believe it shows why "learning" as used in scientific theories of behavior such as Hilgard's bears so little relation to "learning" as used in the classroom. It is not that the behavior-theory concept of learning is just more refined or broken into smaller parts than the classroom concept. It is rather that the rough-and-ready theory of the classroom uses blatantly mentalistic action categories as observables, whereas there is a great deal of resistance to this in behavior theory. Indeed, the very name "behavior" betrays the tendency not to use action categories.

Further, I have tried to point out various implications for research practice as I went along. I have urged careful and continuing attention to the correct description of "what is learned." I have suggested that more consideration be given to theories which claim that human action (intentional behavior) is directly observable. I have pointed out that most behaviorisms, rather than getting closer to the facts in the sense of getting closer to that which can justify our explanations and theories, perversely seem to use categories farther away from the highly reliable ones of common sense. I have urged the necessity for recognizing hierarchies in accounting for learning. I have noted the difficulties surrounding the definitional and empirical use of some of the key terms in learning theory. In this regard I have urged that "reinforcer" as used in the law of effect can have the role of an observable. Finally, I have claimed, especially in the areas of developmental psychology, that the philosophical presuppositions and justifications for the categories in use must surely be displayed in those cases where these categories are themselves being questioned.

In addition to these methodological strictures, let me add a few remarks of a more general nature. More importantly, researchers must stop wasting their time looking for crucial experiments. It should be obvious by now that any theory can be saved from embarrassing empirical consequences if we only add enough "epicycles." But, more profoundly, the crucial experiments tend to be crucial only if conducted within a single research paradigm—that is, only if they operate from the same theory and hence with the same observational categories. For without these common denominators, two theories will just pass each other by through a failure to countenance the same facts.

A corollary of the abandonment of the crucial experiment is the necessity for empirical researchers constantly to be theorizing as well and to be elaborating their theory with its presuppositions as they go. Even if one could argue that this constant appeal to fundamentals is wasteful, at a minimum one must constantly remember that the fundamentals *do* infect ordinary research and be a bit more tolerant of those whose concern is for fundamentals. For the only ultimate way to justify a position and defend it against plausible rivals is by means of a detailed theoretical cum empirical cum philosophical exposition of its own principles and a similarly detailed polemic against the rival positions *in their own terms*. In Kuhn's terminology, only well articulated paradigms can be confirmed or refuted.[38]

Finally, the foregoing gives some support to the thesis that the dominant position in current American psychology—an eclectic functionalism—must be rejected. What I am referring to is a loose attitude of tolerance and borrowing from many traditions while working in a hardheaded way on small localizable problems. Borrowing from many theories without carefully evaluating and incorporating the theory which renders intelligible that which is borrowed can only distort and falsify. The inability to see that some empirical results have the significance they do because of the observational categories and theory in which they are embedded is functionalism's greatest sin. Indeed, the "tolerance" of varying viewpoints seems attractive and fair. But this tolerance is a primrose path shutting out the influence of theory on observation and leading to the stagnation of psychological research. Theorizing is no sin. It cannot be avoided in any event, and claiming that psychology is too immature for theory construction ensures its fixation at that immature level.

Notes

1. Ernest R. Hilgard and Gordon H. Bower, *Theories of Learning*, 3rd ed. (New York: Appleton-Century-Crofts, 1966), p. 9.

   The initial investigation underlying this essay was supported in part by a small project grant from the Office of Education, Contract 0-8-0800233669(010).
2. N. R. Hanson, *Patterns of Discovery* (London: Cambridge University Press, 1958).
3. W. V. O. Quine. *World and Object* (New York: John Wiley & Sons, 1960).
4. Stephan Toulmin, *Foresight and Understanding* (New York: Harper Torch Books, 1961).
5. T. S. Kuhn, *The Structure of Scientific Revolutions* (Chicago: University of Chicago Press, 1962).
6. Hugh G. Petrie, "The Logical Effects of Theory on Observational Categories," Office of Education Contract 0-8-080023-3669(010).
7. R. Rorty, ed., *The Linguistic Turn* (Chicago: U. of Chicago, 1967).
8. Henry Veatch, *Two Logics* (Evanston, III.: Northwestern University Press, 1968).
9. See, for example, George Pitcher, ed., *Truth* (New York: Prentice-Hall, 1964).
10. *Structure of Scientific Revolutions*, chap. 10.
11. *Word and Object*, pp. 73-79.
12. See, for example, I. Lakatos and A. Musgrave, eds., *Criticism and the Growth of Knowledge* (Cambridge: Cambridge University Press, 1970) for a recent discussion of the problem of commensurability as raised by Kuhn's work.
13. W. V. O. Quine, *Ontological Relativity* (New York: Columbia University, 1969), especially the chapter entitled "Epistemology Naturalized."
14. One might also consult Israel Scheffler, *Science and Subjectivity* (Indianapolis: Bobbs-Merrill Co., 1967) for an "opposing" viewpoint which, nevertheless, seems to grant the major points of the theory-dependency thesis.
15. See, for example, G. J. Warnock, ed., *The Philosophy of Perception* (Oxford University Press, 1967).

16. See, for example, K. Koffka, *Principles of Gestalt Psychology* (New York: Harcourt, Brace & World, 1935) and W. Kohler, *Gestalt Psychology* (New York, Liveright, 1947).
17. See also M. Herskovits, D. T. Campbell, and M. Segall, revised by Segall and Campbell, *A Cross-Cultural Study of Perception* (Indianapolis: Bobbs-Merrill Co., 1969).
18. I discuss the problems of operational definitions in much greater detail in "The Logical Effects of Theory on Observational Categories," chap. 3. Also see Hugh G. Petrie, "A Dogma of Operationalism in the Social Sciences," *Philosophy of the Social Sciences 1(1971):* 145-60.
19. See, for example, D. T. Campbell and D. W. Fiske, "Convergent and Discriminant Validation by the Multitrait-Multimethod Matrix," *Psychological Bulletin 56 (1959):* 81-105.
20. Hugh G. Petrie, "The Strategy Sense of 'Methodology'," *Philosophy of Science* 35 (1968): 248-57.
21. The foregoing discussion of the relativity of operational definitions is mirrored in Broudy's discussion of Polanyi's concepts of tacit and focal knowing in the NSSE Yearbook from which the present chapter is taken.
22. R. Rosenthal, *Experimenter Effects in Behavioral Research* (New York: Appleton-Century-Crofts, 1966).
23. Wittgenstein's discussion of "seeing and seeing as" is most illuminating in this regard, *Blue and Brown Books* (New York: Harper Torch Books, 1958); idem, *Philosophical Investigations,* 3d ed. (New York: Macmillan Co., 1968). I have written a paper on this subject which was delivered at the American Philosophical Association, Western Division Convention, May 1970, St. Louis, Missouri. See also my chapter, "Science and Metaphysics: A Wittgensteinian Interpretation," in *Essays on Wittgenstein,* ed. E. D. Klemke (Urbana: University of Illinois, in press).
24. Donald T. Campbell, "Operational Delineation of 'What is Learned' via the Transposition Experiment," *Psychological Review* 61 (1954): 167-74.
25. See Krimerman's contribution to the NSSE Yearbook from which the present chapter is taken, for a theory which uses an action-observation language.
26. Max Weber, *Methodology of the Social Sciences* (Glencoe, Ill.: Free Press, 1949). See chapter by Strike in this volume.
27. The schematic strategy outlined above of showing either question-begging or total implausibility is just the strategy used by N. Chomsky in his celebrated "Review of Skinner's Verbal Behavior," *Language* 35 (1959): 26-58.

Charles Taylor, *The Explanation of Behavior* (New York: Humanities Press, 1964), also mounts a sustained attack on the ability of simple S-R mechanisms to account for complex behavior even in animals. Taylor's major strategy is to show the ad hoc character of the explanations introduced to handle these kinds of embarrassments, to note the experimental implications of these ad hoc explanatory principles, and to point out that where such experiments have been carried out, the ad hoc principles fail in the same kind of way. Since any theory can, logically, be saved by adding epicycle on epicycle, only a thorough critique of this kind can ever discredit a theory. And even then, as Kuhn has pointed out, the discrediting will be effective only when a powerful alternative theory is available. See Thomas Kuhn, *The Structure of Scientific Revolutions* (Chicago: University of Chicago Press, 1962).

28. Wittgenstein, *Philosophical Investigations*, sec. 85.
29. See my detailed discussion of the philosophical issues involved here in "Science and Metaphysics: A Wittgensteinian Interpretation," in E. D. Klemke, *Essays on Wittgenstein* (note 23).
30. James E. McClellan, "B. F. Skinner's Philosophy of Human Nature: A Sympathetic Criticism," in *Psychological Concepts in Education*, ed. B. Paul Komisar and C. B. J. MacMillan (Chicago: Rand McNally & Co., 1967) makes a similar point.
31. Hilgard and Bower, *Theories of Learning*, p. 482.
32. This example is drawn from Donald Davidson, "Actions, Reasons, and Causes," *Journal of Philosophy 55 (1963): 685-790*.
33. I believe this characterization of the law of effect as providing a new set of observation categories is compatible with that given in P. E. Meehl, "On the Circularity of the Law of Effect," *Psychological Bulletin* 47 (1950):
34. See, for example, William Kessen, "Questions for Theory of Cognitive Development," *Child Development Monograph, Serial* 107, 1966, Vol. 31, No. 5, pp. 55-70, for a psychologist's view of such problems.
35. I discuss this problem in more detail in "Science and Metaphysics: A Wittgensteinian Interpretation," see note 23.
36. E.g., Konrad Lorenz, *On Aggression* (New York: Harcourt, Brace & World, 1966); Robert Ardrey, *The Territorial Imperative* (New York: Athenaeum, 1966); Desmond Morris, *The Naked Ape* (New York: McGraw-Hill Book Co., 1968).
37. There is a close connection between what is counted as observable and what must be inferred and what is natural and what must be explained. See Toulmin, *Foresight and Understanding* (note 4).
38. Kuhn, *Structure of Scientific Revolutions*.

[1974]
Action, Perception, and Education

I want to do four things in this essay. First, I want to elaborate just a bit a feedback model of behavior. Second, I will argue negatively that this model is *not* reducible to a straight-line causation analysis of action where this is conceived as a species of a necessary and sufficient conditions form of explanation. Third, I will urge positively that this model captures quite well the important features of common sense teleological explanation. Finally, with reference to two quite different examples—Rosenshine's work on the characteristics of good teachers and Katz's analysis of the role of educational structure as a mediator between social ideology and social change, I will attempt to show how the problems and questions with which they deal are amenable to a most perspicuous formulation in feedback terms.

I

Diagrams are essential in understanding feedback systems. I therefore reproduce the diagram utilized by William Powers in explaining what he calls "the basic control-system unit of behavioral organization."[1]

Despite the fact that Powers uses "stimulus-response" language, it should be clear even from his description that these are stimuli and responses of a very peculiar nature. The "stimulus" or disturbance is only part of what has traditionally been taken to be the stimulus in classical psychology. The other part is supplied by the effects of the organism behaving. Indeed, this is one of the central features of a feedback system—it reacts to its own effects. In fact its effects are "designed" to keep the input quantity as close to the reference level as possible via the action of the effects through the environment on the input quantity.

Furthermore, the S-R laws in Powers' scheme turn out to be laws of the environment—not laws of the organism—once the nature of the controlled quantity is known.[2] The S-R laws are trivial and variability

---

First published in *Educational Theory*, Vol 24, Issue 1, pages 33–45, January 1974

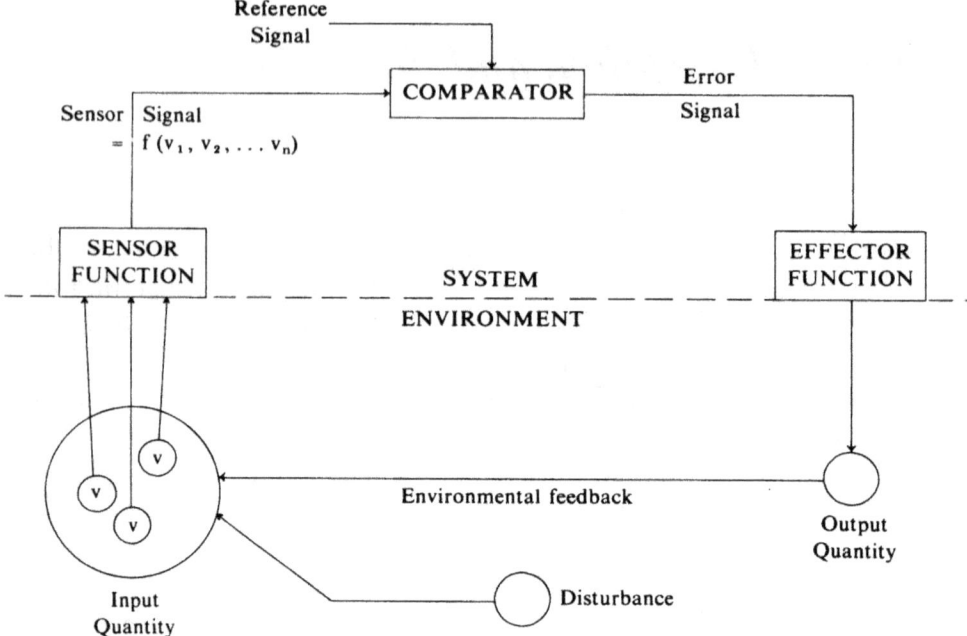

Fig. 1: Basic control-system unit of behavioral organization. The *sensor* function creates an ongoing relationship between some set of environmental physical variables (v's) and a *sensor signal* inside the system, an internal analogue of some external state of affairs. The sensor signal is compared with (subtracted from, in the simplest case) a *reference signal* of unspecified origin. The discrepancy in the form of an *error signal* activates the *effector* function (e.g., a muscle, limb, or subsystem) which in turn produces observable effects in the environment, the *output quantity.* This quantity is a "response" measure.

The environment provides a feedback link from the output quantity to the input quantity, the set of "v's" monitored by the sensor function. The input quantity is also subject, in general, to effects independent of the system's outputs; these are shown as a *disturbance,* also linked to the input quantity by environmental properties. The disturbance corresponds to "stimulus."

The system, above the dashed line, is organized normally so as to maintain the sensor signal at all times nearly equal to the reference signal, even a changing reference signal. In doing so it produces whatever output is required to prevent disturbances from affecting the sensor signal materially. Thus the output quantity becomes primarily a function of the disturbance, while the sensor signal and input quantity become primarily a function of the reference signal originated inside the system.

For all systems organized like this, the "response" to a "stimulus" can be predicted if the stabilized state of the input quantity is known; the "S-R Law" is then a function of environmental properties and scarcely at all of system properties.

of detailed output is seen in a unified way as keeping the input close to the reference level. I will elaborate on this point below, but for now the important thing is to note that this is no ordinary S-R mechanism. Indeed, the explanatory force, as we shall see, goes through the input side of the loop, *not* the output side. Paradoxical as it sounds, what feedback systems *do* is control perceptions—not behavior.

Let us see how this system works in a concrete example. Imagine a central heating (and cooling) system all hooked up to a thermostat, so that the system automatically switches on the furnace when it's too cold, and the air-conditioning when it's too hot. Referring to our diagram, the sensor function is the operation of the thermocouple (or whatever) in the thermostat, and its job is to sense the temperature in the immediate vicinity of the thermostat. This signal representing the actual temperature is compared with the reference signal—the setting on the thermostat. If there is a large enough difference, an error signal of the appropriate sort is sent to the effector function—the furnace and air-conditioning unit. That in turn puts out warm or cool air which, through the environment of the house, affects the temperature which the thermostat senses.

Disturbances typically are the heat loss (or gain) through the house to the outside. However, one could introduce a more drastic disturbance by holding a match near the thermostat, or as my wife and I did once, by putting a lamp near it. It was winter and the house became quite cold because the thermostat was sensing that all was well. This illustrates the earlier point that what the feedback system does is to control its *inputs* where these are a function of *both* disturbances and the environmental feedback *of the effects of the system itself.* It also suggests how we could, knowing the reference signal of 72°—the only intra-system parameter—and the appropriate heat diffusion laws of the house—the environment—*predict* what outputs will occur given what disturbances. Thus, what might have looked like extreme stimulus-response variability to a "behavioral heating analyst," would turn out to be trivially predictable from a knowledge of the controlled quantity of the system—namely the temperature setting. I call my analyst "behavioral" because if he looked only at inputs (house temperatures) and outputs (furnace and air-conditioning behavior), he would effectively be putting the feedback system into a "black box" and looking only at stimuli and responses. He would also observe a great deal of variability, simply by not knowing what the system was *really* doing—controlling the ambient temperature around the thermostat. As soon as he realizes this (as a "methodological" behaviorist, it is likely he never would), the variability disappears.

Furthermore, investigations into "black box" feedback systems to find out empirically what quantity is being controlled are clearly possible. Thus the internal structure postulated by the theory does have empirical implications and can be investigated empirically. Of course, the connection with the world is nothing like the naively direct one enjoined by a behaviorist methodology of operationally defining all internal structural concepts. Thus negative feedback neither relies on mystical purposes nor is it unconnected with the world. It satisfies the requirement for having testable consequences without putting those consequences into the operational definition straight-jacket.

The way it does this is in principle very simple. If one suspects a negative feedback system is in operation, one then hypothesizes a controlled quantity for the system. Note that this "discovery" step depends on intuition and professional hunches, in this case, no more than does the comparable step of suggesting fruitful empirical operations for the behaviorist. Indeed, it is probably because of the close connection between controlled quantities and motives, goals, and purposes in ordinary-language talk about action that common sense provides as many fruitful hypotheses as it does. In any event, once a controlled quantity is hypothesized, the experimental procedure is this: introduce a disturbance near the sensor (it has to be the right order of magnitude so it neither escapes detection nor overwhelms the system) and see if the output opposes the disturbance. If it does, that quantity probably is being controlled. If there is no opposition to the disturbance, the hypothesized quantity is probably not under control. Utilizing the model one can even predict appropriate magnitudes.

In my thermostat example if we hypothesize that temperature around the thermostat is being controlled, we can introduce a disturbance (put a lamp nearby, open a window, etc.), and predict what the temperature change would be near the thermostat if no control system were operating. If the actual temperature change is much less (the air-conditioning counteracts the effects of the nearby lamp), we have hit upon the controlled quantity.

Two more brief remarks need to be made here to give at least a hint of the expansion of such a basic system to account for the complexity of behavior of a human organism. These features will not be needed for the thesis to be developed in the rest of the paper, but omitting mention of them usually detracts from understanding the rest of the system. In the first place organisms almost certainly have hierarchies of control systems with the outputs of the higher order ones probably connected to the reference signals of lower order systems. Thus reference signals can change internally and account for a great deal more, otherwise variable behavior. Furthermore the highest reference signals may be either innately set or changed in the course of learning.[3] I don't think one need appeal to hierarchy or learning in order to understand how negative feedback is radically different from S-R explanation and how it might serve as an explication of ordinary common-sense action explanation; although for a full application of negative feedback to human behavior, such gaps would have to be filled in.

II

My second thesis is that this feedback model of explanation is *not* reducible to an S-R type of model. In a sense this should already be intuitively obvious from my discussion of the behavioral heating analyst. It is true that one *could* record inputs and outputs of the system and attempt to construct some kind of laws, but these would be quite variable, and would "explain" the action of the heating system about as well as behavioral psychologists explain phenomena with their low-level statistical generalizations. Compare this with the explanation afforded by the realization that the system is a negative feedback system designed to reduce the effects of disturbances. (Hence the name, "negative" feedback. Positive feedback in this technical sense sends a system oscillating out of control. Cf. feedback "whine" on a public address system.)

It is, of course, true that there is a causal path which can be followed around the loop of the feedback system. But tracing such a causal path no more explains what the system does on the originally given level than neurophysiology explains action on the psychological level. And this is so even though the psychological and neurophysiological descriptions refer, in some sense, to the same phenomena.

Despite the intuitive plausibility of the foregoing, however, I want to offer a more formal argument against the possibility of giving a necessary and/or sufficient conditions reduction of the concept of explanation by negative feedback.[4] Thus, to the extent that ordinary straight-line causal explanation is a species of necessary and sufficient condition explanation, to that extent will I have shown the impossibility of reducing negative feedback explanation to causal explanation. (It is hard to see from the literature how causal explanation can be viewed otherwise, but I leave the possibility open.)

The locution I will use as paradigmatic is 'X in order that Y.' This is typically teleological language and also fits my model very well. The outputs of the model occur in order that the sensor signal be maintained as close as possible to the reference level given by the reference signal. In order to forestall future misunderstanding, however, let me characterize the range of 'X' and 'Y' a bit more closely. In the first place a 'Y-ish' quantity exists in two places in the system. It exists as a reference signal, a kind of ideal internal representation of a possible external state of affairs. Second the sensor signal is also more or less 'Y-ish' depending on how closely it matches the reference signal. It, however, represents how 'Y-ish' the actual external environment really is rather than some ideal state of that environment as the reference signal does. Thus the thermostat setting of 72°, the reference signal, represents what the temperature should be if the system operates properly. The sensor signal, the signal which represents the actual temperature of the house, may or may not be at 72° at any given moment. In any event, in the explanatory locution 'X in order that Y,' clearly the 'Y' must range over actual states of the environment as represented by the sensor signal; that is, after the actual inputs have undergone whatever transformation occurs in the sensor function. For it would simply make no sense at all to explain the reference signal with the locution, 'X in order that Y.' Whatever is explained by the central heating-cooling system it is *not* the fact that the thermostat is set at 72°. It rather is used normally to explain the temperature of the house.

The situation within the range of 'X' is even more difficult. Primarily 'X' ranges over the outputs of the system, but there are at least two importantly different ways in which these outputs can be described. In the first place one might describe them simply in terms of the operation of the effector function, at least if that is well known. Thus 'X' could be 'the furnace goes on' as well as a host of things bearing a family resemblance to this description, such as, 'the heating system operates,' 'the furnace puts out heat,' etc. The basis of the family resemblance here is the detailed causal operation of the effector function and its immediate effects on the environment.

On the other hand, since the output effects operate through the environment to cancel the effects of any disturbance on the controlled quantity, and since the sensor signal is a function of the disturbance and the output effects, the output can be described in the same "dimensions" as the sensor signal (and, hence, also the reference signal). This is natural enough since whatever other dissipated effects of the output of a negative feedback system, the main importance of the outputs is to change the inputs thereby counteracting disturbances. Thus a different family of descriptions of the outputs would rely on the context provided by the reference and sensor signals—that is, how they represent the world. As an example of this kind of output description one might have 'the house is being heated' as a substitution for 'X'. One might even have 'the temperature of the house is being raised to 72°,' although this might appear to some as not being terribly explanatory.

Thus, the following two examples show the two different kinds of descriptions of output quantities.

a) The furnace is going on in order that the house be maintained at a temperature of 72° F.
b) The house is being heated in order that the house be maintained at a temperature of 72° F.

Notice that in the second example, where X is in the same dimensions as the input quantity, it could be argued that the X is analytically connected with the Y.

This feature of being able to describe output in terms analytically connected with the terms used to describe the controlled quantity (or, to anticipate a bit, the purpose) solves a long-standing problem in the theory of action. Many philosophers have noted an analytic connection between reason or motive or purpose description and the corresponding action description which the reason or motive or purpose is ordinarily taken to explain.[5] They have concluded, mistakenly, that since there is this analytic connection, there cannot be a causal connection. With the aid of the feedback model we can see that at least some descriptions of output will be analytically connected with descriptions of the reference level since the very nature of the system is to use outputs to counteract disturbances of input from the level of control set by the reference level. Yet there is nothing operating around the loop but ordinary causal mechanisms.

With these preliminaries out of the way, let me turn to the more formal argument. If 'X in order that Y' is to be used as an ordinary explanatory schema, then presumably X must be either a necessary or sufficient condition of Y or both. But X cannot be a sufficient condition of Y. For if it were, then to the extent that Y admits of alteration in magnitude then alterations in X will produce proportional alterations in Y. Thus if a rise in demand is sufficient to cause a rise in supply, a large rise in demand causes a larger rise in supply. But as I have shown, precisely the opposite effect occurs in a feedback loop. Variations in X are utilized to *maintain* Y in its reference condition. Thus either heating or cooling outputs maintain a temperature of 72°.

Second, Y need not actually occur for a feedback explanation to be appropriate and true. (Recall that this is 'Y' in the sense of the sensor signal representing the actual state of the environment.) Yet if Y does not occur where X is thought to be a sufficient condition of Y, X cannot explain Y. Thus it may be so cold outside that the temperature cannot, given the efficiency of the furnace and the insulation of the house, be brought to 72° even if the furnace runs continuously. Nevertheless, it would be true to say that the furnace ran in order that the house be maintained at a temperature of 72°.

But perhaps most importantly of all, assuming X to be a sufficient condition for Y simply reverses the ordinary sense of what is being explained by what in a locution like 'X in order that Y.' Commonly, we would take Y to be explaining X in such a case rather than X explaining Y as the sufficient condition supposition seems to require.

Perhaps then, X is a necessary condition of Y in 'X in order that Y'? Well, perhaps in *some* sense, but not in any ordinary causal sense. For to repeat an earlier point, in a feedback explanation, Y need not actually occur for the explanation to be true. Yet if we remember that *what* we are explaining is X's occurrence, we would be faced with the unpleasant prospect of having, in ordinary causal terms, to explain the occurrence of X without the occurrence of Y, X being a necessary condition of Y. This is indeed mysterious on straight-line causal analysis, but perfectly straightforward with a feedback model.

In addition, on at least many analyses of necessary and sufficient conditions, if X is a necessary condition of Y, then Y is a sufficient condition of X. But not only is it true, as I have already shown, that Y need not occur for a feedback explanation to be true, but it is also true that Y can occur in a feedback system without X's occurring. This would be impossible if Y were a sufficient condition of X. Thus if the house just happens to be at 72°, then the reference condition is maintained although no heating system output occurs. The system, however, "stands ready" in a clear sense to maintain its goal. Could this be similar to the sense in which according to Ryle in motive or dispositional explanations a person "stands ready" to overtly behave in the appropriate way?[6] As you may have guessed, I think so.

III

In this section 1 want to argue that the basic feedback model fits very well our ordinary notions of action and what would be required for action explanation. There are probably about as many accounts of action explanation as there are philosophers of action. However, there are some things upon which they seem to agree and it is my hope that in arguing for the compatibility of the feedback model with one of these accounts, I will be showing the possibility of utilizing the feedback model in an explication of our ordinary notions of action.

For this reason I have chosen the account of action explanation offered by Charles Taylor in *The Explanation of Behavior*.[7]

Essentially Taylor urges that action explanation is a special form of teleological explanation. In turn, teleological explanation which is paradigmatically concerned with such locutions as 'X for the sake of Y' has at least the following characteristics:

1) In teleological explanation, the observed order or pattern is itself a factor in its own production.

2) The explanatory scheme must not be "metaphysical" in the sense of postulating some unobservable entity, which can make no difference in the observed world.

3) The explanation must be irreducible to non-teleological explanation.

4) The natural tendencies or purposes of the system must be irreducible to explanation by any other more basic laws of a different form. That is, the teleological explanation must be basic for the normal operation of the system. Taylor calls this the asymmetry of explanation.

What must be added to teleological explanation to get action explanation according to Taylor is that in action explanation, the action to be explained must be "directed to a goal" or, very generally, intentional. In a bit more detail, this seems to include the following:

1) The behavior to qualify as action must be describable in terms not merely of a goal or end-state actually reached, but rather in terms of a goal or end-state posited or intended by the organism. The organism must have a view of the world which guides its action independently of whether or not that "intention" actually corresponds to the world. In common terms, this is just a philosopher's way of pointing out that people act in terms of their perceived world, whether or not that corresponds to reality.

2) But, moreover, not only must the organism have such an intention, that intention must somehow be operative in producing the action.

3) And, yet, paradoxically the role of the intention in producing the action cannot be causal, at least in any straightforward sense of 'causal'. Philosophers of action tend to fall back on the notion of 'agency' as a primitive notion to explain these otherwise contradictory requirements of the intention's producing the action, yet in a non-causal way. But, of course, naming the mystery does *not* explain it, and the one great advance of the feedback account is that it *does* offer an explanation for this otherwise opaque notion.

4) Action can, at least sometimes, be affected by an appeal to reasons, desires, motives, etc., and such an appeal shows that in accounting for action an appeal to reasons is justificatory rather than explanatory in a standard sense. "Reasons are not causes" as the saying goes.

How well does the negative feedback model fit into this analysis of action explanation? Let us see. Beginning with the teleological requirement that the observed order is in part responsible for its own production, that clearly occurs in the feedback model. Indeed I urged above that the inputs in an operating feedback system are almost wholly determined by its outputs. So in a feedback system maintaining its sensor signal at the reference level the pattern created by the outputs is the pattern sensed; this pattern is then compared with the reference signal which produces the error signal which drives the output. The pattern operates in its own production.

As for the necessity for empirical import, I have already indicated the general procedure for testing whether one is dealing with a negative feedback system with an hypothesized controlled quality: introduce a disturbance and see if it is counteracted by the outputs. Furthermore, the whole of the last section was spent in arguing for the irreducibility of the 'X in order that Y' locution to a necessary or sufficient condition form of explanation. Furthermore, I tried to suggest that such a form of explanation *is* appropriate to the negative feedback model. Recall the discussion of the heating behaviorist?

Nevertheless, some things probably still nag, for we do believe that in some sense the actual detailed operation of say, the heating system, is subject to ordinary causal laws. Doesn't that indicate that the feedback model, at least in this case, really is reducible after all and hence is *not* a form of teleological explanation? I think not. The notion of reducibility is not at all clear and repeated attempts by philosophers of science to clarify it have not been terribly successful. In any event one can reject the present argument even without having a clear notion of reducibility. For if the mere fact that thermostats do operate in accordance with mechanical and electrical laws is sufficient to show that a putative teleological explanation of their behavior is reducible to the electro-mechanical laws, then likewise the mere fact that human beings obviously have a physiological basis for their actions is sufficient to show the reducibility of action explanation to physiological explanation. But this is thought to be absurd by a lot of people, even including behavioral psychologists who also do not believe in any ghosts in the machine. Surely "the furnace went on in order to heat the house to 72°" is a perfectly acceptable explanation of its kind, and in a fairly clear sense *what* it explains is not at all what would be explained by a detailed tracing through of the electro-mechanical causal chain.

But this problem can be attacked in another way if one recalls that what is controlled in a feedback system is the perceived quantity. There are no reference signals delicately controlling the detailed outputs. Indeed the reference signal can stay the same and the detailed outputs can vary considerably precisely to counteract the effects of disturbances on the controlled quantity. Feedback controls *sensed* quantitites, *not* outputs. It affects outputs, to be sure, but it does not control them. This term 'control' is a technical notion which refers to the operation of the feedback system to maintain the sensed signal near the reference signal no matter what the disturbance. The perceived quantity will match the reference signal by means of a wide range of outputs. Recall the furnace again. The signals from the thermostat do affect the furnace outputs, but knowing that will *not* tell one how many BTU's the furnace will put out in any given hour. The reason is that if we introduce a disturbance (an open window) into what the thermostat senses, then *that* is what the system tries to counteract.

One might suggest that the system also controls the outputs since if a burner on the furnace gets plugged (a disturbance), the furnace will just run longer to put out the same amount of heat.[8] This is true enough as far as it goes, but there is a clear sense in which this kind of control is in reality an indirect consequence of the actual feedback control. For what determines the amount of heat the furnace must put out in a given time in the first place? The feedback system controlling, in the technical sense, the sensed temperature. Further, over a period of time there will be considerable variability in the amount of heat put out by the furnace on different occasions of operation. Yet all this variability is used by the system to control (in the technical sense) the sensor signal. The point is crucial and *not* widely recognized. Feedback systems control perceptions, not outputs. (Indeed, this fact gives an insight into understanding the gestalt nature of perception in a non-mysterious way—but that is another story.)

How does this control of perception notion help to understand the feedback model as a form of teleological explanation? Well, in ordinary causal explanation the "cause" *does* affect the outputs (the effect) but in feedback the "cause," the system of sensor signal, reference signal, and error signal controls the inputs. In ordinary causal chains disturbances intervening between cause and effect change the effects. In feedback systems, disturbances are opposed so that the effect—the sensory signal staying near its reference level—stays the *same*. In this sense, then, the operation of feedback systems is clearly *not* reducible to a straight-line causal analysis.

Furthermore, control of perception by feedback systems also gives a sense to the required asymmetry of explanation in teleological explanation. What typically is not explained by causal laws is the ordinary functioning of the system. What *can* be explained by causal laws is when the system goes haywire. In this sense the normal or natural tendencies are not explained by causal analysis, although abnormalities can be so explained. The current went off, the furnace ran out of fuel, etc.

Let me turn then to the added features which are supposed by Taylor to turn teleological explanation into action explanation. First was the requirement that the behavior must be intentionally describable in terms of goals or purposes. This somewhat puzzling and controversial requirement in standard action theory becomes crystal clear under the feedback model. If what the system does is to control the perceptual signal, making it match the reference signal, then clearly the behavior is describable in terms of the goal, the purpose, the reference signal. The outputs (behavior) can also be described in other ways, but those may not be nearly as perspicuous for understanding the operation of the system. The furnace puts out so many BTU's per hour, but why? To maintain a temperature of 72°. The organism's muscles contract, but why? To sink that putt. If feedback systems control perception, then they operate on the world as experienced from the "inside."

But the intention must also, in action theory, not only be present; it must produce the action without causing it. This pair of requirements has caused no end of trouble for action explanation. On the one hand, if intentions really cause actions, then as the behaviorists point out, an intention does so by means of being a link in a causal chain which link occurs inside an organism and is itself, in its turn, caused. Hence any special function for an intention would appear superfluous. In brief, explanation by intention would not be teleological after all. And yet ordinary language analysis of action explanation shows it to be teleological through and through. Is such a basic part of our conceptual scheme just so much left-over mysticism to be dispelled by the advances of behaviorism? Given the non-advances of behaviorism, this has not seemed so to many.

On the other hand, if intentions and reasons and purposes do not cause action, then why are they necessary? A spirited defense of the thesis that reasons *are* causes has been offered by Donald Davidson in his very influential paper, "Actions, Reasons, and Causes."[9] Davidson has convincing arguments that reasons cannot possibly explain actions unless they cause them—after all there may be reasons which give a rationale for an action without thereby necessarily explaining the action—somehow the reasons must actually *be* the ones which are operative in the given case. If the mode of operation is not causal, it remains mysterious. Despite this approach, however, the spectre of reductionism has haunted action theorists if once they let causation in the door.

What is needed is a model which shows how reasons could be causes *without* permitting reductionistic explanation. The feedback model is just such an answer. The operation of the model around the feedback loop is clearly causal. Yet, if my preceding arguments have been at all effective, it should be clear by now that the mode of explanation appropriate to the feedback model is *not* reducible to the linear causal model. The causal operation goes around the loop from input through comparator to output through the environment and back to the input. The explanation runs in the *opposite* direction, from reference signal through perceptual signal to the environmental situation as *intentionally* characterized. Action theory can have its cake and eat it too. Intentions, reasons, etc., *can* cause action *without* falling prey to reductionism.

If reasons can be causes, then what happens, finally, to the longstanding distinction between explaining and justifying? In the first place if reasons are not reducible to causes, as I have argued, then justification in terms of reasons is not obviously reducible to explanation in terms of causes. The distinction would seem to survive Whatever the nature of the activity of justification turns out to be, it is not obviously threatened by viewing action in terms of the feedback model. Without then attempting to explain justification in terms of feedback, at least in this paper, let me offer one or two observations which might begin to hint at what is involved. Taylor speaks of justification, or the giving of reasons as having, at least occasionally, an effect on the purposes, goals, and reasons of another. These kinds of entitites are generally to be understood as reference signals in a feedback model. Further, an adequate model for an organism would involve a hierarchy of such basic units as I have sketched here with the outputs from higher orders serving as reference signals for lower orders. Thus, if there is a level concerned with reasons, a reference signal could represent being a good reason, the perception of a given reason could be compared with the reference level which in turn could effect outputs thereby changing lower order reference signals and in turn changing overall behavior. Reasons, and the fact that they are good reasons, sometimes do effect behavioral changes. Furthermore, the reasons, in order to effect this change, must be perceived *as* reasons and *not* just as elements in the causal loop.

In short, I conclude that conceiving the feedback model as an explication of ordinary action and as a basis for action explanation is fully justified. The clear features of our intuitive account of action are captured, some puzzles in the ordinary account are resolved, and the remaining difficulties concerning reduction and the justification-explanation distinction do not seem insurmountable.

IV

In conclusion, I will briefly consider two educational problems to which the foregoing account can be turned. In a survey of the literature on teacher enthusiasm, Barak Rosenshine distinguished between high-inference and low-inference measures of teacher enthusiasm.[10] High-inference measures are those which require a lot of "inference" from the observation of the behavior to the labeling of the behavior as, e.g., energetic, expressive, and so on. Low-inference measures are much easier to classify, e.g., words per minute, teacher requests opinion, and so on. Further, although both kinds of measures seem positively correlated with pupil achievement, Rosenshine concentrates on one study in which teachers kept the content and organization of a lesson the same and simply manipulated the animated behavior (low-inference). This by itself increased pupil achievement, and so Rosenshine suggests that one possibility for increasing teacher effectiveness would be for colleges of education to aid prospective teachers in becoming enthusiastic or even in feigning enthusiasm, for that seemed to work just as well! Surely if that's all research on teaching has been able to turn up, something has gone radically wrong somewhere.

The feedback model can, however, help us out here. Under this model what good teachers do is try to control their perceptions that good education is taking place. If they sense an error they act in such a way as to *remove* the error. Furthermore, as the central heating system example shows, a feedback system can emit outputs which are diametrically opposed to one another if described solely at the level of the outputs. The furnace at one point puts out so many BTU's per hour. At another time it takes so many BTU's per hour *out* of the house. Yet under the proper "intentional" description, both of those outputs are describable as the *same* thing—maintaining the temperature of the house at 72°. Feedback systems *vary* their outputs in order to counteract the effects of disturbances on what the system is controlling.

Now suppose that whatever good teaching is, it is represented by a reference signal *in* the teacher. If there is an error between that reference signal and the perceptual signal of teaching coming in, the error signal will operate the output function and in some situations, on the level of the outputs, exactly contrary outputs can both be instances of good teaching. The difficulty with attempting to characterize Rosenshine's high-inference measures by means of varying combinations of low-inference measures (observable behaviors) is just this: if the low-inference measures of good teaching used in the studies summarized by Rosenshine were on the level of the outputs, then it is highly likely that they would simply cancel each other out when subjected to standard statistical treatments. This typically leads psychologists to be very suspicious of the high-inference measures and to concentrate on the behavioral outcomes. For example, intuitively we know a good teacher should sometimes praise and sometimes correct a pupil's mistake. Yet if these two teacher characteristics were tested in standard ways, they would almost surely cancel each other out.

The typical psychologist's response in such a situation would be: "Well let's specify the characteristics of the situations in which the teacher should praise and those in which the teacher should correct the student." The problem here is that such specification is just not possible except on the level of "those situations which call for praise *in order to* teach effectively." The teleological notion creeps in.

The neat thing about feedback is that it doesn't *care* what the situation or environment is in detail. It only cares if the sensor signal matches the reference signal and if not, it will generate outputs which will in turn affect the sensed signal. If the system fits the environment at all well, it can handle an indefinite number of detailed variations within that environment by means of an indefinite number of different detailed outputs—without separately commanding each of those outputs. Behavior is the control of perception.

Indeed this analysis applies generally to the problem of operational definitions. They are usually wrong-headed because they tend to define outcomes on a level lower than what the feedback system is controlling. And yet the "higher-order" goals are, as I have shown, open, at least in principle, to empirical investigation. Hypothesize a controlled quantity, introduce a disturbance, and see if it is counteracted.

My second educational example comes from the history of education. In *Class, Bureaucracy, and Schools,* Michael Katz argues that educational structures mediate between ideology and the results of schooling.[11] Furthermore, this mediation is an absolutely essential one in the sense that one could not change the goals of schooling without changing the structures. Thus, in a sense he is claiming that the *function* of the educational bureaucracy is to promote class bias and racism, where "function" here is a term often used by sociologists to indicate some stronger kind of relationship than mere consequence.

And, of course, Katz needs the stronger relation to make out his case. For if the class bias and racism is merely a consequence of the system, no matter how unwanted, it remains conceptually possible to intervene in the ordinary operations of the system to block these consequences. It would then be theoretically possible to change the consequences *without* changing the system. And, indeed, just this kind of criticism is typically leveled against radical critics like Katz. On the other hand, if the racism is a function of the system, then perhaps Katz can maintain his stronger thesis.

Typically, however, sociological methodologists have been totally unable to explicate the notion of the "function of a system" in any way which avoids reduction to a straight-line causal analysis. Thus, although anthropologists may claim that a certain rite of passage performs a certain function in a society, it is not clear what more this means than that the puberty rite has the purported function as a consequence. Similarly one can grant the evidence Katz cites as to racism in American education but claim this is nothing more than an undesirable consequence, to be eliminated by intervention in the straight-line causal sequence which gives rise to the racism.

Feedback theory, however, provides the conceptual apparatus to find a way out of this impasse. It would take only a moment's reflection to see that the functional explanation beloved of sociologists, if it is to be anything more than straight-line causal analysis, must be interpreted in the social sphere analogously to teleological explanation in the psychological sphere. Thus what makes a sociological explanation a truly functional explanation is the existence of reference level for a controlled quantity with the system organized so that its operation will tend to counteract perceived disturbances in the controlled quantity.

Thus if Katz believes that a function of schooling in this sense is to promote the social status quo, and hence to limit social mobility, the crucial kinds of evidence he needs are of the following sort: find cases in which it looks as if the schools ought to be seen as promoting social mobility, then check to see if the school system itself reacts to counter this disturbance. Thus the inability of society to absorb over-educated young people, and the recent attempts to foist a manpower notion on schools, would seem to be evidence in favor of Katz's hypothesis as would the renewed attacks on the native intelligence of blacks just at a time when they are beginning to gain access to "establishment" higher education. On the other hand, the single most potent source of criticism of society continues to be the universities. It is still the case that nowhere else can a radical get as influential a hearing.

Of course my purpose here is not to judge the truth or falsity of Katz's thesis, but only to point out the kind of argument he needs if he is ever to convince the liberal establishment of the thoroughgoingness of his critique. He must not merely build up evidence of class bias as a consequence of schooling, no matter how widespread. Rather he must look at those crucial cases wherein there would appear to be a disturbance in the hypothesized controlled quantity and see if the system acts to counteract that disturbance.

It would also prove most suggestive to look at the educational system with feedback categories in mind. What, for example, serves the perceptual input function? Is it boards of education, teacher-training institutions, teachers themselves, or what? In brief, what mechanisms provide the perceptions of reality in the system called "schooling"? Next, what functions as the comparator? Very likely this is a feedback formulation of "Who has the power?" The nature of the output function also may be suggestive. What outputs really do correct perceived disturbances? Is it administrators of individual schools or might it be researchers urging the heritability of I.Q. on us to counteract the disturbances of low-class blacks demanding a part of education?

In summary, I have argued that feedback theory provides a relatively detailed model of how to understand some of the most vexing problems in action explanation and sociological explanation. It provides a hardheaded alternative to a behaviorism which is already intellectually bankrupt while at the same time removing some of the "mystery" surrounding more traditional formulations of humanistic insights. It could provide an individualistic learning and teaching model which could supplant the current reliance on statistics of groups. For no matter how much one is reminded that many statistical laws apply only to groups, the temptation is well-nigh irresistible to apply the statistics to individual members of the group leading to the inhumane kind of depersonalization so much decried in our schools of late.

Notes

1. I am indebted to William T. Powers, *Behavior: The Control of Perception* (Chicago: Aldine, forthcoming, 1973) for the basic notion of feedback control. The diagram and accompanying description is from William T. Powers, "Feedback: Beyond Behaviorism," *Science,* 179, 26 (January 1973), pp. 351-356.
2. See William T. Powers, "A Feedback Model for Behavior: Application to a Rat Experiment," *Behavioral Science,* 16 (1971), pp. 558-563.
3. Such matters are discussed most suggestively in Powers, *Behavior: The Control of Perception, op. cit.*
4. I am indebted for the main lines of this argument to Alan Ryan, *The Philosophy of the Social Sciences* (New York: Pantheon, 1970), Ch. 8.
5. See, for example, A. I. Melden, *Free Action* (London: Routledge and Kegan Paul, 1961); R. S. Peters, *The Concept of Motivation* (London: Routledge and Kegan Paul, 1958), and classically Gilbert Ryle, *The Concept of Mind* (London: Hutchinson, 1949) among others.
6. Ryle, *op. cit.*
7. Charles Taylor, *The Explanation of Behavior* (London: Routledge and Kegan Paul, 1965), Chs. 1-3.
8. I owe this observation to Robert H. Ennis.
9. Donald Davidson, "Actions, Reasons, and Causes," *Journal of Philosophy,* LX, No. 23 (November 7, 1963), pp. 685-700.
10. Barak Rosenshine, "Enthusiastic Teaching: A Research Review," *School Review,* 72 (1970), pp. 499-514.
11. Michael Katz, *Class, Bureaucracy, and Schooling* (NY: Praeger, 1971).

[1975]
Can Education Find Its Lost Objectives Under the Street Lamp of Behaviorism?

There is no need to document the extent to which behavioral objectives (or some variant thereof) have taken over in education. Colleges of education are urged to institute plans for performance-based teacher education. State departments of public instruction demand that school systems institute schemes for writing behavioral or instructional objectives. Private firms have been contracting to increase levels of student "performance," where such performance is usually stated as a certain level of test scores. Legislatures and school boards, faced with constricting financial resources, demand that education be made accountable, from institutions through teachers, for the public funds allotted to them. And underlying all these phenomena is the recurring theme of behavioral objectives as a universal panacea.

Yet the more closely one looks at these varying manifestations of the increased influence of that peculiar social scientific doctrine known as behaviorism, the more one is struck by how out of phase are the ills being attacked with the cures being offered. One cannot help but be reminded of the story of the drunk crawling around on his hands and knees under the street lamp. A passerby notices this odd behavior and asks, "Lose something?" "My watch," replies the drunk. The well-intentioned samaritan proceeds to help the drunk look with no success. Finally, he asks, "Just where did you lose your watch?" "Over in that doorway," replies the drunk, pointing down the street a hundred feet or so. "Well, why in the world are we looking for the watch here, then?" explodes the samaritan. "Oh, the light's better here," comes the reply.

It would appear that if one could show that the current emphasis on behavioral objectives is as out of touch with reality as the drunk's behavior, we would not have to await the empirical evidence to know that education will not be able to find its way under the street lamp of behaviorism. It is my contention that there are so many conceptual confusions rampant in applying the doctrines of behavioral objectives to education

---

First published in: Smith, R. (ed.) Regaining educational leadership: essays critical of PBTE/CBTE. New York: Wiley.

that no matter how bright the light, it is falling on inappropriate ground. In this paper, I want to isolate some of the more important of these confusions and attempt to show both the justifiable kernel of truth along with the overlay of misunderstanding and error. My hope is that once these confusions are laid bare, the individual institution, principal, teacher, student, or layman can intelligently separate the wheat from the chaff.

I

One of the central confusions in this general area is the conflation of accountability with measurement—and a peculiar kind of measurement at that, specifically, the measurement of efficiency. One cannot seriously challenge, for example, the request from legislators that schools must be held accountable for their contribution, or lack thereof, to the public good. Indeed, hard though it may be to swallow, there is nothing inherently wrong about a society asking itself whether it values welfare for its poor more than education, or decent housing more than either, or if nuclear deterrence is the highest value of all. Social values do not come automatically ranked and in an economy of scarce resources; it is surely plausible that sometimes hard decisions must be made.

Schooling as a social institution will have to continue to make its case as best it can—sometimes in competition with other social goods. The one change I would urge is that more effort be made to justify the products of schooling, for example, education, as good in itself rather than solely on the basis of schooling's instrumental value to other goods. The opposite kind of justification has been the more prevalent. During the 1960s one could scarcely pick up a Sunday newspaper supplement without reading an article on how much more money college graduates would make than nongraduates, or on how desirable liberal arts graduates were in business, never mind their major. This line of reasoning clearly holds up schooling as an instrumental value; it is a means to other ends. It is small wonder, then, that when the bubble burst in the early seventies and schooling was no longer a means to a better job, support for schooling fell off sharply. Had schools paid even a modicum of attention to justifying the intrinsic value of education, they would not look so stupid now—they would have some kind of intellectual ground to fall back on to justify their ultimate value.

The point, however, is that it is surely acceptable to hold schools accountable for their contribution to the public good. This sense of accountability means nothing more than holding responsible in the typically ethical sense. We hold government responsible in this sense for the rapprochement with China and for Watergate. We hold industry and technology responsible in this sense for increasing our standard of living as well as for polluting our environment. Schools are no different in this respect.

There is another sense of "accountability," however, which is all too easily confused with the above sense. This second sense is the one from which the profession of accounting is derived. In an almost literal way it has come to mean to count up or measure. And it is most important to notice here that for this "counting" sense of accountability to get a handhold, the *units* in terms of which we count must be externally determined. Accountability in the measurement sense cannot tell us *what* to measure. Thus, this kind of accountability can only show us how efficiently we are pursuing already agreed-upon goals. It cannot evaluate between alternative goals. It is when accountability in the "holding responsible" sense is confused with accountability in the "measurement" sense that mischief results.

These two senses of accountable can be illustrated quite nicely by considering the case of industry. As long as there is general agreement on profit as a goal of industry and no serious perceived conflict between that goal and other social goals, there is no difficulty in holding industry accountable by measuring how they generate profit. Accountants perform this measurement task by means of financial statements, and a good measure of the efficiency of the business is given by these statements. But when a question is raised as to the desirability of the profit motive as a single all-inclusive value, for instance, when questions of accountability in the sense of ethical responsibility are raised, as in the case of pollution, it simply makes no sense to look at financial statements to find out if business is accountable. A question of goals themselves is being asked, *not* a question of the efficiency with which already agreed-upon goals are being met. Financial statements can, of course, reflect pollution fines and the cost of antipollution measures. What they cannot do is give any answer to whether or not industries ought to be held accountable for their pollution.

Now, on almost anyone's showing, education's goals include as a major part the development of the skill, sensitivity, and intellectual curiosity to reexamine traditionally accepted goals. There is no doubt some sense in which progress toward such goals can be determined, but "efficiency" models of measurement are precisely the wrong ways. For efficiency models depend on agreed goals, and the goals of educators are to question these agreed-upon goals. It scarcely makes sense to measure the progress of some activity by means of accounting procedures whose very reliability is derived from what the activity is questioning. The way in which accountability is presently being practiced in education is like allowing the men accused in the Watergate scandal to determine the rules for whether they are guilty. You might accidentally get justice, but it would not be very likely.

Thus, in discussions of accountability, educators must be perfectly willing to morally justify the institution of schooling as an important social good with all the intelligence and fervor at their command. At the same time, however, they must demand that the nature of education renders "efficiency" models of measurement wholly inapplicable as guides to whether schools are doing their jobs.

II

Even if one is clear that the efficiency model of measurement is inappropriate to accountability in the schools, another confusion is rampant in discussions over how one does find out if schools are doing their jobs. This is found in its most general form in the confusion between the desirability of competence as a measure of learning and the actual state of the art of evaluation. In more specific terms it is urged, for example, that what we want are competent teachers and not necessarily people who have gone through some specialized curriculum. Thus we get the push for performance-based teacher education, the use of instructional objectives in the classroom and what have you with a corresponding increase of attention on assessing these competencies.

Now it is clearly an advance conceptually to recognize the distinction between some desired competency on the one hand, and perhaps any number of ways of attaining that competency on the other hand. And it is a good thing for schools occasionally to be reminded that their cherished courses, lectures, and recitations may not be the only way to achieve the competencies aimed at in education. On the other hand, it must be remembered that a standard school curriculum clearly is *one* way of achieving educational competencies.

Nevertheless, this very valuable conceptual distinction between end product and varying ways of getting to the end product is not automatically translatable into practice. If one is going to aim at and certify for competencies actually possessed rather than curricula undergone, then clearly one must be able to judge in some reliable way when the competency actually is present. Thus, the empirical side to the competency-based coin is an adequate, comprehensive, and reliable method of assessing the presence or absence of the competencies. Yet virtually everyone, even those who are in the testing establishment, know that current testing models are really extremely unreliable. Rough distinctions can be made, but current testing procedures cannot even come close to making the fine-grained distinctions that are required if we are, as a matter of policy, to abandon curriculum satisfaction in favor of competency-based criteria for having achieved educational goals.

Let me take just one example of the inability of current tests to make the fine-grained distinctions needed. Scores on the prestigious College Entrance Examination Board (CEEB) tests determine for many students whether or not they will go to a certain college. Yet by the CEEB's own admission, there is only a 68% chance that a score difference of 31 points on the verbal test (34 points on the quantitative) represents a real difference in ability.[1]

Thus, if that difference occurs around the college's cutoff point, one student will be admitted and another not. Of course, college admissions people will be quick to say that they use other indicators as well. It turns out that gradepoint average, clearly a curriculum-based indication, is *the* most reliable indicator of college success. But this just makes my point for me. Judging competencies without taking into account curricula undergone is very, very difficult.

Harry Broudy makes this point in an instructive way.[2] One occasionally finds someone practicing medicine, apparently very successfully, who has never been to medical school. Does the AMA admit that he has the competency to be a doctor and let it go at that? Not at all! But that is just what schools of education are being urged to do. The point is that it may just be that given the present and foreseeable state of the testing art, the very best, although not infallible, way we have of judging whether someone possesses a specific competency is whether he has undergone a standard school curriculum in that area.

Thus, educators should quickly grant the distinction between possessing a given competency and various means by which the competency might have been attained. But until the testing establishment demonstrates *its* competency to detect competencies a *great* deal better than it can now, the prudent response would be to go very slowly in converting to competency-based programs. A good strategy would be to agree with the principle of such programs and then examine very carefully the detailed testing arrangements to be used.

III

Examining the details of testing programs almost inevitably leads one into another one of the confusions rampant in the educational use of behavioral objectives. This confusion is between the desirability of having some sort of connection between an ascribed competency and the real world on the one hand, and a particular specified kind of connection on the other hand.

It is a truism that any scientific theory must have some sort of empirical import. And likewise, if a supposed educational goal, whether it be a competency or any other kind of educational result, has no conceivable connection with the world as we can observe and experience it, then clearly such a goal is some kind of chimera. Therefore, some kind of publicly testable result surely is a necessary condition of any supposed educational outcome, and the testing establishment is certainly on unassailable grounds in insisting that there be *something* we could observe—sometime, somewhere.

Nevertheless, this truism from the philosophy of science is confused with the idea that the required observability be of a very peculiar behavioristic sort. Behaviorism in general holds as a methodological principle that the only kinds of things that can be observed in the world of human beings are very gross kinds of "behavior," and that all else must be inferred. It turns out, however, that this principle is honored more in the breach than in the observance. If it were strictly followed, behaviorism would have been seen to be impossible long ago. Only constant equivocation on its own methodology gives it even its remote plausibility.

For, if we are to follow behaviorism, we can observe marks on a piece of paper, but we must infer test results and competence. We cannot observe that a student understands the material. We can see a student standing in front of one painting longer than another, but we cannot see that he appreciates one more than another. We can see a student moving around in his seat with knitted brow, but we cannot see that he is confused.

I have argued in technical detail elsewhere that this view of what is observable and what is not is dogmatic and totally unsupported by any argument or evidence.[3] However, the point I wish to make here is that it is extremely poor policy to confuse the necessity for testability of some form or other with a highly controversial specification of just how that testability is to be understood. Public policy in the schools cannot possibly justify *requiring* everyone to implicitly adopt one side of an obviously controversial issue in the methodology of the social sciences. This is especially true since the acceptance of testability does *not* also require the acceptance of behaviorist language and methodology.

Thus, educators should grant the necessity for their educational goals to "make a difference"—to be testable somehow, somewhere. They should be clear and precise about these goals and desired outcomes, but they should also insist on the necessity for the tests to be appropriate to *their* goals and not vice versa. And where there is an inconsistency between available testing methods and the professed goals of experienced practitioners in any field, it is at least as likely that testers need to be more imaginative as it is likely that the practitioners have not known what they have been doing all that time. In short, educators must insist on the priority of the goals in determining the appropriateness of the tests.

IV

A fourth confusion has been recently expounded in great detail by Michael Scriven.[4] However, it is so important as to bear summarizing here. It is desirable to be objective (reliable) in one's judgments about educational as well as all other matters. Contrariwise, one should attempt to avoid insofar as possible being subjective (biased). Unfortunately, the pair of terms "subjective" and "objective" have another set of meanings that are often confused with the "biased—reliable" set. Subjective also means relating to feelings, thoughts, emotions, and judgments of a *single* person. Objective as a contrast to *this* sense of subjective simply means intersubjective, that is, referring to the feelings, thoughts, emotions, and judgments of *more* than one person.

Now clearly one can be subjective in this second sense without at all being biased or unreliable. The most obvious example is that any person is usually the most reliable judge of his own internal states of emotion and, for that matter, of his own thoughts. But there are educational examples as well. Indeed, graduate education is based upon the belief that subjective (personal) judgments of the student's adviser are more reliable than the judgments of lots of other people. Even the fact that there are usually doctoral *committees* does not vitiate this point, since one must pick people for the committees who are *qualified*. And how do we know if they are qualified? Are they qualified if their subjective (personal) judgments are reliable and unbiased?

Another example lies in the area of art appreciation. The subjective (personal) judgment of a sensitive art teacher as to the progress and competence of a student in coming to appreciate art is almost certainly far more reliable than any intersubjectively verified test. As a related example, any sane person would surely prefer Leonard Bernstein's subjective (personal) judgment as to the quality of a student violinist to the satisfaction of so-called "objective" behavioral objectives. Nor did Bernstein attain his eminence by satisfying behavioral objectives.

And yet the confusion is so rampant that not only behavioral objectives buffs, but many others as well would prefer an objective intersubjectively verified test to a subjective judgment every time—even in areas where we *know* that the objective measurement is not as reliable as the subjective judgment. The results of this confusion are to be seen in the dreary sameness of our schools. The intersubjective tends to reduce everything to the lowest common denominator. Even worse are the effects that a denial of the reliability of subjective judgments has upon individual students. We are faced with students being labeled as mentally retarded and largely condemned to a certain kind of education largely independent of subjective judgments of their ability. And yet the push for behavioral objectives can only intensify the denigration of subjective judgment and the tendency to replace reliability with intersubjectivity, whether such replacement can be justified or not.

Thus, the educator should embrace reliability and shun bias. But he should be very careful to understand in each case wherein reliability resides. Quite clearly it is not always in majority opinion. Traditionally the judgment of trained experts has been considered most reliable in some areas. When asked to replace that judgment by something "more objective," the educator should very carefully ask whether reliability is likely to be increased or whether results will simply be homogenized to the benefit of no one except the "objectifiers."

V

The next confusion is closely related to the last one between subjective and biased and objective and reliable. It also arises in discussions emphasizing student competence as over against teacher performance and so is related to the earlier discussion of competence and curriculum and the difficulties of determining competence independently of curriculum. This present confusion is between an emphasis on student outcomes and competences as desirable educational goals on the one hand and appropriate ways of assessing teacher performance on the other hand.

Surely *what* the student learns should be a major part of any set of educational goals. The goal of education is *not* for a teacher to go through a rigid lesson plan independently of whether or not students learn anything from that lesson plan. The teacher's performance must be relevant to the student's learning in some sense or other. Obviously it would be senseless to suppose that teaching performance has no connection whatever with student learning performance.

On the other hand, it seems equally obvious that teacher performance cannot be wholly, or perhaps even mainly, judged on the basis of student outcomes.[5] There are simply too many factors other than the teacher's performance that go to determine the student's ultimate performance. The student may be lazy and not learn from the brightest teacher. He may be highly motivated and learn from the dullest teacher. The student may not have the competence to learn, or he may be so bright that he learns from *any* teacher. Are teachers in high socioeconomic neighborhoods that good, or is their job so much easier? Are ghetto teachers all bad or are they contending with virtually insuperable environmental problems? In short, there are clearly many occasions on which the best teaching efforts will fail and many other occasions on which very poor teaching may, nevertheless, be associated with good student learning.

Consider an analogy from baseball. One of the goals of the shortstop is to commit as few errors as possible. Indeed, one could not understand someone "playing shortstop" and being totally unconcerned with the number of his errors. However, there can be very bad shortstops who commit very few errors. They never try very hard so they never get close enough to the ball to commit many errors. On the other hand there can be very good shortstops who commit many errors. They are trying for everything. Evaluation of good shortstops cannot be tied too closely to one outcome.

Analogously, one can imagine situations in which the very best teaching would result in no student learning of what was taught. Consider the case of the student who in terms of interest or ability simply should not be taking a certain course. The best teaching in such a situation would result in getting the student *out* of the course and hence in his not learning anything about the course at all. Therefore good teaching is and must be kept conceptually distinct from actual student learning.

Thus the educator should grant that one of the goals of teaching may well be student learning. Occasionally teachers forget this and seem to think schools would be great places if only there were not any students. They should not be allowed to forget their responsibilities to students. But, likewise, neither should judgments of their professional competence be judged solely or even mainly on their student's performances. Schemes of teacher evaluation should be developed that rely on student progress to an *appropriate* degree but that also have provision for significant weight to be given to professional, peer, and self-evaluation.

VI

The last confusion is one that really should not have to be mentioned. Unfortunate as it is, proponents of many of the accountability-type programs discussed in this paper seem unable to distinguish between criticism of the concepts they use and criticisms of the people who employ these concepts. Time after time, when it is pointed out that behavioristic concepts have this or that kind of implication, the response is that the practitioners do not actually do such nasty things. In logic this response constitutes the fallacy of *ignoratio elenchi*—missing the point.

A person who is committed to any set of principles is also logically committed to the logical implications of those principles. If he at the same time does not wish to be committed to those implications, then he is simply inconsistent and, of course, literally anything and everything logically follows from inconsistent premises. Is it any wonder that accountabilists tend to be so hard to pin down? If one is allowed to take contradictory stands, then one is bound to be right half the time *and* bound to be wrong the other half, and, worse, one cannot tell which is which.

The saddest thing, however, is that the accountabilists seem to feel no necessity to respond to the criticisms offered of their programs other than to say they do not behave in the undesirable ways indicated by the logical consequences of their principles. Now it may well be that by acting inconsistently with their own principles, accountabilists can avoid, for a time, the implications of those principles. But ideas have a way of catching up with inconsistent uses. If the ideas are mistaken or confused, the action taken in the name of those ideas will likewise in the long run be mistaken or confused. History allows no more lenient interpretation.

Thus, the educator should continue to point out the intellectual confusions, where they exist, to the accountabilist. And if the accountabilist responds as if he were being personally attacked, the educator must gently, but firmly, point out to the accountabilist that he has simply missed the point of the criticism. The truth or falsity of ideas is sometimes hard to grasp, but false ideas cannot stand the light of reason forever.

VII

My initial analogy between the drunk looking for his watch under the streetlight and the use of behaviorism in many areas of education can now be seen to have been quite generous to behaviorism. To fully reflect the confusions I have tried to illustrate in this paper the story would have to run something like this: The drunk would have to be very pleased with himself if he elaborately and *efficiently* covered the ground under the street light even without finding his watch (the confusion between accountability and measurement of efficiency). Furthermore, the drunk should find a child's toy watch, be unable to tell the difference between that one and his own, and still be perfectly satisfied. After all, the end product was achieved, finding a watch (the confusion between concentrating on the end product and the sorry state of our ability to determine whether the end product has actually been achieved). The drunk's original behavior of looking under the street light for the watch because the light was better there, even though he knew the watch was not lost in that area, would be retained (the confusion between needing observability for an empirical theory and thinking it must be behaviorist observability). The drunk ought to be looking for a "time" he had forgotten rather than a watch, but still be visually looking under the street light because others might be able to help him in such a public "objective" search (the confusion of intersubjectivity with reliability and bias with personal feelings or thoughts). The drunk would have to feel that he, the drunk, was a superb searcher if the samaritan found the watch and gave it to the drunk (the confusion between good teaching and student learning as the sole criterion of good teaching). Finally, if the samaritan were to criticize any or all of the drunk's ideas, the drunk would have to get mad and claim that he really does not act absurdly (the confusion between the implications of ideas and behavior inconsistent with those ideas).

If all of these changes were made to the original story, one would get a very confused drunk. One would also have a very good analogy with the current situation of accountability and its cousins in education. It really is that bad.

Notes

1 *College Boards Guide* for *High Schools and Colleges,* 1972-1973 (College Entrance Examination Board, 1972).
2 Harry S. Broudy, *The Real World of the Public Schools* (New York: Harcourt Brace Jovanovich, 1972), p. 65.
3 Hugh G. Petrie, "A Dogma of Operationalism in the Social Sciences," *Philosophy of Social Science I* (1971), pp. 145-160.
4 Michael Scriven, "Objectivity and Subjectivity in Educational Research," *Philosophical Redirections of Educational Research,* Seventy-first Yearbook of the National Society for the Study of Education (Chicago: University of Chicago, 1972).
5 See Israel Scheffler, *The Language of Education* (Springfield, Ill.: Charles C Thomas, 1960), Chap. 2, for a classical statement of the issues surrounding the conceptual connection between teaching and learning.

[1974–76]
Do You See What I See?
The Epistemology of Interdisciplinary Inquiry

It seems to me that since the answer to the question in the title of my paper is, for members of interdisciplinary groups, not always and obviously, yes, an examination of why this is so and how it might be overcome is in order. The impetus, and, indeed, part of the content for this investigation arose out of my participation in the Sloan Program of the College of Engineering at the University of Illinois over the past two years. That program was in large part designed as an interdisciplinary effort to examine the role of the social sciences and humanities in an engineering curriculum. The method was interdisciplinary faculty seminars, and my particular interest was in the processes which occurred in those seminars. I was a general participant in the meetings which brought in a series of speaker-discussants on the topics, "How does X View the World." "X" was each week replaced by the name of the discipline of the speaker. In addition, I chaired an interdisciplinary subgroup whose topic was the interdisciplinary research and teaching process. Much of what I will say in the following is a result of these experiences, and although a philosopher, I will be making some non-philosophical claims in what follows. That I dare to do this is part of what must result, I think, if interdisciplinary work is ever to be successful.

Harry Broudy has surely been one of the most persistent advocates, at least of late, for the importance, not to say the necessity, of interdisciplinary work.[1] Basically the argument is that a complex technological

---

This paper was presented at a three-day interdisciplinary conference "The Uses of Knowledge in Personal Life and Professional Practice" held at the University of Illinois at Urbana-Champaign in September, 1974, upon the occasion of Professor Harry S. Broudy's retirement as professor of philosophy of education from the University of Illinois. First published in: *Educational Researcher*, 1976, 5(2), 9-15. —Republished (1976) in *The Journal of Aesthetic Education*, 10(1). 29-43. —Reprinted (1986) in Chubin, D. E., Porter, A. L., Rossini, F. A., and Connolly, T. (eds.), *Interdisciplinary analysis and research*. Mt. Airy, Maryland: Lomand. 115-130.

society requires interdisciplinary solutions to its problems. And I think the argument requires little restating. One need only consider the problems of pollution, world-wide inflation, energy production and conservation, and so on to get the flavor. In addition if one adds the increased sensitivity of professional schools to their broader social roles as evidenced by other papers, the importance of interdisciplinary work becomes apparent.

Unfortunately, the importance of interdisciplinary work has seldom been matched by its fruitfulness. All too often grandly conceived interdisciplinary projects never get off the ground and the level of scholarship seldom exceeds that of a glorified bull session. All too frequently, people look upon interdisciplinary projects as a dumping ground for the less than disciplinarily competent—and justifiably so. Yet, as I shall argue, it is only from among the most competent disciplinarians that an interdisciplinary group can draw its members if it hopes for success. It is in hopes of contributing to a higher rate of success for interdisciplinary projects that I offer the following "profile" of interdisciplinary inquiry—research or teaching.

First, however, a few preliminary distinctions need to be noted. I distinguish between interdisciplinary and multidisciplinary efforts. The line is not hard and fast, but roughly it is that multidisciplinary projects simply require everyone to do their own thing together with little or no necessity for any one participant to be aware of any other participant's work. Perhaps a project director or manager is needed to glue the final product together, but the pieces are fairly clearly of disciplinary size and shape. Interdisciplinary efforts, on the other hand, require more or less integration and even modification of the disciplinary subcontributions while the inquiry is proceeding. Different participants need to take into account the contributions of their fellows in order to make their own contribution.

Take the energy crisis, for example. If the heating engineer as a member of a group looking at energy consumption in housing is simply asked to design houses which are more thermally efficient, he can do that in an almost wholly disciplinary way. He needn't worry about energy cost structures or legal restriction, etc. On the other hand, if the group is considering significant changes in social organization and life-style to meet the energy crisis, the same engineer will have to take projected altered living styles and arrangements, different patterns of energy consumption, and so on into account in order to do even his disciplinary work. And, of course, conversely with respect to the non-engineering participants. It is the interdisciplinary as opposed to the multidisciplinary process with which I shall be concerned in this paper.

The other distinction I want to make here is that I shall not be concerned with the single person who acquires more than one disciplinary competence. In the first place, such a person's problems will be mirrored, I think, in what I shall say about the workings of interdisciplinary groups. But second, such a solution, given the demands of time and energy placed on attaining even one disciplinary competence, is simply out of the question for most people. If we cannot stop short of making Renaissance persons out of a good deal of our population, then the interdisciplinary mode will *not* be able to contribute to the solutions of our pressing societal problems. Thus, I shall talk about groups instead of individuals.

With these preliminaries out of the way, let me indicate briefly what I shall be doing in the remainder of my paper. First I shall note several very important nonepistemological factors which seem to be particularly relevant to the success or failure of interdisciplinary inquiry. These include the notion of idea dominance, psychological considerations, and the institutional setting in which interdisciplinary work is carried on. Next, I shall turn to the epistemological and methodological constraints on interdisciplinary work. Here I shall concentrate on the problems raised by the apparent fact that different disciplines utilize different observational categories and occasionally mean quite different things by the same linguistic terms. I shall suggest that the kind of knowledge exhibited by knowing the observational categories and meanings of the key terms of any discipline is fairly close to what Broudy calls the interpretive use of knowledge. I shall then expand on this notion of interpretive knowledge as a universally necessary condition for successful interdisciplinary inquiry. I shall indicate how one can, in principle, tell when it has been obtained, and I shall conclude by noting the key pedagogical concept necessary for coming to understand the language of a wholly different discipline, viz., metaphor.

Nonepistemological considerations

Idea Dominance. One of the central considerations necessary for interdisciplinary success seems to be what I will call the dominance of an idea. That is, there must be a clear and recognizable idea which can serve as a central focus for the work. It can be embodied in a single individual who leads the project through force of personality or importance of the perceived mission. The dominant idea may be imposed by some external necessity clearly perceived by all participants. Certain kinds of mission-oriented projects fit here. Finally, it may be an idea embodied in a new and powerful theoretical concept or model—a concept which does not find a natural home in an established discipline.

Closely associated with the idea dominance is the necessity for some kind of achievement. The need for achievement also appears under the heading of psychological characteristics, but here it is primarily directed toward the logical requirement of some kind of feedback to confirm the clarity and force of the idea originally conceived. Thus a dominant personality begins to lose dominance if the group cannot be led to some sort of achievement. A mission unachieved raises doubts as to whether it was properly defined. And a powerful new theoretical idea will ultimately be shelved if it fails to achieve results.

The notion of idea dominance seems to admit of degrees—there can be more or less of it. I would predict that, other things being equal, the stronger the idea, the more chance of success. A caveat must be entered here. In some cases it may be extremely difficult, if not impossible, to judge the strength and dominance of an idea independently of whether it turns out to be successful or not. However, at least a gross empirical handle does seem possible here in that this criterion would seem to rule out interdisciplinary projects undertaken simply for the sake of being interdisciplinary.

Psychological Characteristics of Participants. The second major category of nonepistemological factors is the psychological characteristics of the participants in a successful interdisciplinary effort. Of course, successful people here are very much like successful people in any endeavor, but several characteristics, attitudes, and motivations stand out. The person must, first of all, be secure in his or her original endeavors. Interdisciplinary efforts seldom work if the participants are not fully competent in their own fields. Second, the participants must have a taste for adventure into the unknown and unfamiliar, i.e., they must not be tied too closely to their secure home base. Of course, there is a sense in which a really good disciplinarian is, ipso facto, adventurous. It is a taste for *new* adventure that I am talking about here. Third, their interests must be fairly broad, if not in terms of their spheres of competence, at least in terms of what they feel is of importance.

It should be noted that disciplinary competence and security are sometimes at odds with broad interests and imaginative speculation. Given the current pattern of graduate education, the kind of people attracted to any discipline will tend to be those who are good at a fairly narrow thing. Furthermore, the rewards to a successful academic tend to reinforce the narrow, albeit incisive, disciplinary focus. Thus, on the whole one tends to see good disciplinarians uninterested in interdisciplinary efforts and many who are interested seem to have marginal disciplinary competence. A useful blend of competence and broad interest is rare.

The need for achievement enters into the psychological realm as well. Not only must there be achievement in the sense of the development and confirmation of the dominant idea as already mentioned, the participants must also feel that they are achieving something. This need magnifies the difficulties of combining in one person security of disciplinary competence and broad interest, for the external signs of achievement upon which people depend generally do not go to the person interested in interdisciplinary work. Thus the ability to get internal satisfaction and a sense of achievement are crucial in the early stages of interdisciplinary work. Providing signs of achievement might also be a very effective way for administrators and those concerned with the social setting of the effort to protect the very fragile nature of interdisciplinary projects in their early stages. (See the discussion of the sociological setting below.)

The precise mix of disciplinary competence, adventurous spirit, and broad interests may be very difficult to determine. What does seem clear is that no one of these can be allowed to predominate. By this I do not mean that the extremely competent disciplinarian would not make a good participant in an interdisciplinary effort, but rather that if he is not also extremely adventurous and extremely interested in the project, the rewards which accrue simply due to one's disciplinary competence are likely to pull one away from the interdisciplinary effort. Likewise, the person of extremely broad interests, but lesser disciplinary talent may feel the project is going well, but in fact it never gets beyond the superficial. And the adventurous spirit is needed for learning, where necessary, parts of new disciplines.

Another set of psychological issues involve the simple dynamics of working in a group. It has been remarked over and over by members of Sloan subgroups that they seem to spend almost all semester simply learning what each other is like, everyone getting their biases and interests on the table, and only after this is done do they feel they could really get to work. Whether or not they really could get to work is not at issue here. What is important is that such a "shakedown" of attitudes and modes of behavior is almost always necessary with new groups before they can get to a more substantive level of functioning.

The Institutional Setting. This third category involves the institutional setting for the interdisciplinary work. Under this head is included first of all administrative support for the project. This involves seed money, released time, encouragement, and so on. Closely related to administrative support is the necessity for peer recognition somewhere. This can come from the original parent guild of the participant, from a larger community which deems the interdisciplinary work important, or from the interdisciplinary group itself. These features are connected with the achievement need mentioned under idea dominance and psychological characteristics.

Thus I would predict generally that the more administrative and social support which can be given to interdisciplinary groups, the more successful they are likely to be. Complicating the situation, however, is the need to recognize the operation of the dynamics of the group. With very strong idea dominance, some of the early settling in may be avoided; but in the main, one will simply have to be realistic about how much can be accomplished in a given time under conditions in which the members of the group are, almost by definition, strange to each other.

Epistemological considerations

I turn now to the category of epistemological considerations. This general area is involved with the modes of inquiry appropriate both to the parent disciplines of participants and to the interdisciplinary effort. In the first place the participants need to recognize that different disciplines do have different cognitive maps and that these maps may well get in the way of successful interdisciplinary inquiry.[2] By cognitive map here I mean the whole cognitive and perceptual apparatus utilized by any given discipline. This includes, but is not limited to, basic concepts, modes of inquiry, what counts as a problem, observational categories, representation techniques, standards of proof, types of explanation, and general ideals of what constitutes the discipline. Perhaps the most striking of these, and also often the least noted, is the extent to which disciplinary categories of observation are theory and discipline relative. Quite literally, two opposing disciplinarians can look at the same thing and not see the same thing.[3] I hope to illustrate this thesis in a few moments.

The present point, however, is that if disciplines do differ in their cognitive maps, then quite plainly until these maps are shared by the interdisciplinary participants, they will be unable to see the relevance of their colleagues' points of view to the problem at hand. If they do not learn the other disciplinary maps, at least some of the discussion will be necessarily misunderstood for it will be processed in terms of the participant's *own* map which may not be the same as that of the person who offered the comment in the first place. Thus learning at least a part of other disciplinary maps is a necessary condition for turning multidisciplinary work into interdisciplinary work.

It might be objected here that learning another discipline's cognitive map cannot possibly be a logically necessary condition for successful interdisciplinary work, since we can point to numerous cases in which the nature of the problem itself clearly called for the insights afforded by another discipline.[4] Thus at a certain stage in the development of biology, the problems clearly called for the insights of physics. The examples could be multiplied. Of course, I do not deny the existence of such historical examples. What I do wish to dispute, however, is that there really was no learning of the cognitive maps of the other relevant discipline. After all, not all biologists saw the need for physics. Could it be that those who did had already learned the necessary minimum about physics?

Alternatively, I would imagine there are cases where people believed that the insights of another discipline were relevant to their current problems, and yet upon investigation and greater familiarity with the other discipline they found that their early faith was misplaced. History seldom records such failures, but they would seem to indicate that problems "call" for other disciplines only when enough is known of the other disciplines to make the call *appropriate*. In short, my claim that learning (or having learned) at least a part of other disciplinary maps is a necessary condition for interdisciplinary work is a conceptual rather than an historical claim.

I would also hypothesize that a failure to undertake such learning helps explain the relatively naive character of so much interdisciplinary work. Failing to realize the significant differences in cognitive maps and yet faced with the necessity for communicating with each other on *some* level or other, the participants retreat to the level of common sense which *is* shared by all. But ipso facto, such a level cannot make use of the more powerful insights of the disciplines. On the other hand some very successful interdisciplinary work has occurred because the overlap of cognitive maps was large to begin with as, for example, in nuclear engineering or biophysics. The problem is paramount when the maps are far apart as, for example, when the team involves humanists and scientists.

Given this difference of cognitive maps, the question arises of what kind of learning of another's disciplinary map is required for the interdisciplinary team member. Harry Broudy gives us a clue here. He has distinguished four uses of learning—the associative, the replicative, the applicative, and the interpretive.[5] Roughly these uses of learning are as follows. One uses learning associatively when on the occasion of use, the learning provides a context of associations. Aesthetic learning in the appreciation of art often functions associatively. One uses learning replicatively when one replicates the learning on the occasion of use in just the form in which it was learned. Spelling is a prime example of replication. One uses learning applicatively when one *does* something in light of the learning. A great deal of expertise is required here both to know the theory, how to apply it, and when to apply it. Persons exercising their full disciplinary competence probably are using their learning applicatively. Finally, one uses learning interpretively when the situation of use is interpreted with the aid of the learning. It is *seen* in light of the learning.

Broudy also suggests that the interpretive uses of learning or knowledge should be understood primarily in terms of Polanyi's concept of tacit knowing.[6] I cannot even begin to do justice here to Polanyi's rich and fertile discussions of tacit knowing. It will be sufficient for my purposes to note two things. First, I take my development in the remainder of this paper to be in the spirit of Broudy's interpretive use of learning and Polanyi's notion of tacit knowing.

Second, I shall make direct use of one central feature of tacit knowing: the contrast between tacit and focal knowing as that is exemplified by the Gestaltist's figure-ground relationship. Polanyi's claim is roughly that the figure in perception is known focally while the ground is known tacitly. Furthermore, as one shifts to perceiving the ground focally, the former figure recedes into the ground and becomes tacit. It is clear that tacit knowing would prove to be a valuable addition to Broudy's theory of interpretive uses of knowledge. For if the interpretation is tacit it would explain both the importance of the interpretive use as well as the difficulty of justifying that use in an age in which everything seems to have to be made focal in behaviorist terms in order to be recognized as important.

Tacit knowledge used interpretively can also be seen as extremely suggestive for my problem of how much and what kind of the others' disciplines must be learned for successful interdisciplinary work. One needs to learn enough so that this knowledge can be used to interpret the problem in the other disciplinary categories. Interpretive knowledge is almost surely used tacitly by the disciplinarian and this explains why it is so easy to overlook its importance in interdisciplinary work. Further, one often retreats to a common sense which is tacitly used by all when the going gets rough. My claim is that one *can* and probably must make this interpretive knowledge focal so that all can learn it well enough to enable it to function tacitly from then on in the operation of the group.

The minimal constituents of the amount of learning needed of the others' disciplines seem to be the following: First one must learn the observational categories of the other discipline and, second, one must learn the meanings of the key terms in the other discipline. Note that this would seem to allow one to interpret the problem in the others' terms but stops short of the full-fledged knowledge of theory, modes of inquiry, and ideals of the discipline, which the disciplinarian himself would possess. It would allow one, however, to understand the import of certain claims or recommendations made from the disciplinary point of view. Such knowledge by the participants in an interdisciplinary group is certainly not sufficient for success, but as I have argued, it is necessary and, clearly, has been largely overlooked in the past.

Let me try now to illustrate what I mean by observational categories and meanings of key concepts. Consider the following so-called "ambiguous figures."7

Do you see the martini in the first figure? Now, how about the torso of the girl wearing a bikini. Do you see the duck-rabbit in the second figure? Now consider the third figure of the young-old woman. This one is hard for many. The old woman is looking down and to the left. The young woman is looking away from the viewer and to the left. The old woman's mouth is the choker around the young woman's neck. Notice how, as Polanyi claims, what is focal for one interpretation becomes tacit for another. Note too how the cognitive concept seems to give meaning to the lines or parts of the drawings rather than the other way around. Imagine what it would be like to be a member of an interdisciplinary group discussing a problem in which the young-old woman, or something analogous, played a part. What would happen if your discipline allowed you to see the young woman, while another's discipline interpreted it as an old woman and you didn't realize the difference? Would you be tempted to retreat to your own narrow discipline and categorize those other folks as just silly? My suggestion here is quite simple. It often happens that when different disciplines look at the same thing (the same lines on the paper) they *observe* different things. Thus, it is necessary for people engaged in interdisciplinary work to understand each others' observational categories.

The second example concerns different meanings of key concepts and comes from a discussion section of a course I teach. I was sitting in on the section as an observer and the teaching assistant was trying to explain the difference between facts and values. He gave as an example, "Blacks score ten to fifteen points below whites on standard I.Q. tests," and asked whether this was a statement of fact or a statement of value. A classmember responded that it was a statement of value. This was *not* the correct answer. As discussion proceeded, it became clear that a very understandable difference of meaning was being attached to the concepts "fact" and "value" by the student and by the T.A.

By "fact" the T.A. meant any statement which *purported* to describe what is the case, whether we know if it is true or not. Thus controversial claims and even false claims were all facts to him at least as opposed to values which purport to say what *ought* to be the case. For the student, fact was limited to true, noncontroversial facts and all else was value. Again if different disciplines have different meanings for the same terms and this is not taken into account, one can predict almost certain failure for interdisciplinary projects.

But now if the interpretive use of tacit knowledge is what is required in the interdisciplinary situation, almost by definition it will be a difficult task to determine when the appropriate knowledge of observational categories and theoretical meanings will have been attained—at least short of full disciplinary training. This problem is particularly vexing when one considers just how systematically ambiguous varying interpretations of the world might be among several disciplines. Think of the young-old woman again. Two different disciplinarians might talk about "the woman" which both of them see for a long time without realizing they were talking about different things. For a long time some very intelligent people thought, and perhaps some still do think, that former President Nixon was really talking about a humanitarian response to Hunt's plight rather than hush money on that infamous March 21 tape.

The solution to this problem of how to tell when someone has learned a set of observational categories and theoretical meanings is in principle deceptively simple: introduce what would be a disturbance into the situation being observed and see if the other person counteracts the disturbance.[8] If the disturbance *is* counteracted, the appropriate categories and meanings probably have been learned. What does this mean? Consider again the young-old woman. If one is attempting to determine whether someone has learned to see the old woman, one might suggest that despite her age, she certainly has a lovely nose. If one can actually see the old woman, that should, for most, constitute a disturbance which would be resisted by some disclaimer as, "You call *that* nose lovely? You're out of your mind." In the case of the teaching assistant and the student who disagree on the meaning of "fact," the assistant can introduce examples of true facts, false facts, and controversial facts to see what sort of resistance the student puts forth. If all of these count as facts while a paradigmatic value statement is not, and vice versa, then the student probably understands the fact-value distinction.

A real life example occurred once in one of our general sessions during the Sloan project. Professor Nicholas Britsky was speaking to us on how the artist views the world and was showing us a series of slides of his own and others' work. Recall that we were a thoroughly interdisciplinary group with scientists, engineers, social scientists, and humanists. Professor Britsky came to a slide of one of his own abstract works. He was asked whether a certain predominant color area on the canvas could have been anywhere else, and his response was negative. Some in the audience agreed. To move that area would have constituted a disturbance from the perspective of the observational categories of those who understood and appreciated the art work. Others could not see the difference that would have been made. They had not yet assimilated the appropriate observational categories. The principle of introducing a disturbance to test for the presence of categorical and meaning knowledge is thus clear even if the application is often extremely difficult.

I have now identified the minimal cognitive level necessary for successful interdisciplinary work—namely, coming to use the observational categories and meanings of the other disciplines interpretively. I have also indicated, in principle, a test for when this use of knowledge has been attained. In conclusion, I want to sketch briefly the key pedagogical tool which I think needs to be employed to bring people to this minimal level of understanding of another's discipline. The tool I have in mind is metaphor—where "metaphor" is conceived of broadly as encompassing visual metaphors and even theories—models as they are often called in the sciences.[9]

Notice that the interdisciplinary situation is, by hypothesis, one which seems peculiarly apt for the kind of language which has surrounded metaphor.[10] The participants are familiar with one set of observational categories and meanings, their own, and they want to gain an insight into another system of observational categories and meanings. Metaphors traditionally have enabled us to gain an insight into a new area by juxtaposing language and concepts familiar in one area with a new area. One begins to see the similarities and differences between the literal uses of the metaphor and the new area to which we have been invited to apply the "lens" or "cognitive maps" supplied by the metaphor.

The notion of correcting disturbances enters again into the actual pedagogical use of metaphor. The students in the group begin by utilizing the inferences, concepts, and observational categories surrounding the literal use of the metaphorical term in the new to-be-learned area. Of course, certain adjustments are made due to the dissimilarities already perceived by the student between old and new areas of discourse. However, since the learning is being conducted in the presence of an already competent disciplinarian, the student who makes a wrong move with the metaphor creates a disturbance for the disciplinarian teacher. The disciplinarian's reaction shows the student that the move under discussion is part of the difference between the literal use of the metaphor and the new use. Gradually both come to react to disturbances in the same way, as already described.

I cannot begin to give many varied illustrations of pedagogically useful metaphors, primarily because it follows from my discussion that only competent disciplinarians can locate their own best metaphors. However, I shall try to give at least two. My own presentation here has used the ambiguous figures as visual metaphors for the important notion of theory-dependent observational categories. I have found by experience that this metaphor is usually extremely good pedagogically.

A second example I still remember from high school geometry. A very dear, old-fashioned teacher used it to explain the concepts of point as location, line as distance, and rectangle as plane surface. She held up a pencil and said, "Imagine this pencil sharpened as sharp as possible—and then sharpened much sharper than that. That's a point." Then she took this "point" in her fingers and drew it apart, saying, "Now, if I take the point and draw it apart like this, that's a line." Then she pulled the line down in front of her saying, "And if I pull this line down like so, that's a plane." For me that metaphor worked beautifully, and I think most disciplinarians would be able to come up with appropriate pedagogical metaphors for their own fields.

An important pedagogical point here is that through use and assimilation, metaphors die and take on simply an alternative technical meaning. When disciplinarians fail to realize that terms which they use in a technical sense—as dead metaphors—may be taken as quite live metaphors by their students, communication problems are almost certain to result. Thus the conscious and imaginative use of appropriate metaphorical devices seems to be required to bring the members of an interdisciplinary group to the requisite minimal level of understanding of each other's discipline. Once more we return to one of Harry Broudy's long-standing interests—the importance of humanistic education in general and, to the extent that metaphor is central to aesthetic education, to aesthetic education in particular. Although in this case, I'm not quite sure that Professor Broudy will approve of *that* much stretching of aesthetic education.

Summarizing, I have argued that for truly interdisciplinary as opposed to multidisciplinary efforts, the factors of idea dominance, psychological characteristics of the participants, and the institutional setting are all extremely important. With respect to epistemological considerations, I have urged that some mixes of disciplines require as a necessary condition for success that the participants must learn the observational categories and meanings of key terms of each other's discipline. This knowledge is then tacitly used in an interpretive way on the problems facing the group. One can tell when this minimal learning has been achieved by noting when disturbances are corrected. Finally, I have suggested that a conscious attention to a very broad notion of metaphor is the key of bridging the gap between the differing categories and concepts of the different disciplines. Only when you see what I see does interdisciplinary work have a chance.

Notes

1. See, for example, Harry S. Broudy, *The Real World of the Public Schools* (*New* York: Harcourt, Brace, Jovanovich, 1972), Ch. 7, and Harry S. Broudy, "On Knowing With," *Proceedings of the Philosophy of Education Society* (Edwardsville, Ill.: Studies in Philosophy and Education, Southern Illinois University, 1970), pp. 89-103.

2. Thomas Kuhn's work is probably the best known current position on the differences in cognition among different disciplines. See Thomas Kuhn, *The Structure of Scientific Revolutions*, 2nd ed. (Chicago: University of Chicago, 1970).
3. There is a large literature on the theory-dependence of observation. A classical source is N. R. Hanson, *Patterns of Discovery* (Cambridge: Cambridge University Press, 1958). A view which accepts much of the theory-dependency thesis yet objects to some of the more radical interpretations of it can be found in Israel Scheffler, *Science and Subjectivity* (New York: Bobbs-Merrill, 1967). For some of the pedagogical implications of this view, one might consult Hugh G. Petrie, "The Believing in Seeing," in *Theories for Teaching*, ed. by Lindley J. Stiles (New York: Dodd-Mead, 1974).
4. This problem was suggested to me by Dudley Shapere.
5. See Harry S. Broudy, B. O. Smith, and J. R. Burnett, *Democracy and Excellence in American Secondary Education* (Chicago: Rand McNally, 1964).
6. See Broudy, "On Knowing With," *Philosophy of Education Society*; Michael Polanyi, *Personal Knowledge* (Chicago: University of Chicago, 1958); and "The Logic of Tacit Inference," *Philosophy*, Vol. 40 (1966), 369-86, will get one started on Polanyi's views of tacit knowledge.
7. The martini-bikini was drawn for this paper. The duck-rabbit and the young-old woman were taken from N. R. Hanson, *Perception and Discovery*, ed. Willard E. Humphreys (San Francisco: Freeman, Cooper and Co., 1969), p. 90.
8. This notion of a disturbance and counteracting a disturbance which I am here using in what is hoped to be a nontechnical way receives a most illuminating and far-reaching technical treatment in William T. Powers, *Behavior: The Control of Perception* (Chicago: Aldine, 1973). In ten years this book will have generated a revolution in philosophy and psychology.
9. See my paper, "Metaphorical Models of Mastery: Or How to Learn to Do the Problems at the End of the Chapter in the Physics Text," presented to the Philosophy of Science Association meeting, Notre Dame, November, 1974, for a detailed analysis of the role of metaphor in pedagogical situations logically equivalent to the one obtaining for interdisciplinary work. This paper is scheduled for publication in *PSA 74: Boston Studies in the Philosophy of Science*, Vol. 32, 1975. See also Andrew Ortony, "Why Metaphors Are Necessary and Not Just Nice," *Educational Theory*, Vol. 25, No. 1 (Winter 1975), 45-53; and Felicity Haynes, "Metaphor as Interactive," *Educational Theory*, Vol. 25, No. 3 (Summer 1975), 272-277.
10. The account of metaphor upon which I am relying is a fairly standard one as found, for example, in Max Black, *Models and Metaphors* (Ithaca, N.Y.: Cornell University, 1962).

[1976]
Metaphorical Models of Mastery: Or, How to Learn to Do the Problems at the End of the Chapter of the Physics Textbook

Without question, one of the most important cluster of issues in recent philosophy of science has centered around the attack on the rigid positivist distinction between theory and observation or between a theoretical language and an observational language. Kuhn, Feyerabend, Hanson, Toulmin, and Polanyi are all names closely associated with one version or another of this attack.[1] All have argued that observational categories are essentially theory-determined and there is no determinate observational base, or neutral observational language. Thus at least the positivist account of the objectivity of scientific knowledge would seem to be seriously threatened by the thesis of the theory-ladenness of observation. For without an independently accessible observational base against which to test scientific theories, wherein would objectivity consist?

A number of philosophers of science have rallied to the defense of objectivity against the threat posed by the thesis of the theory-ladenness of observation. One of the earliest defenses and still one of the most reasonable and persuasive was offered by Israel Scheffler in his book, *Science and Subjectivity.*[2] Scheffler by no means defends a phenomenalist or positivist account of observation or objectivity, but rather grants a good deal of the thesis of the theory-ladenness of observation. His strategy is to attempt to save objectivity while granting that observation is essentially theory or cognitive-laden.

Whether Scheffler is successful or not in this attempt will not be my concern here; indeed that controversy still rages. Rather I shall attempt to show that the philosophical thesis of the theory-ladenness of observation even in the attenuated form accepted by those such as Scheffler provides an extremely difficult problem for science education. Indeed,

---

R. S. Cohen et al. (eds.), Philosophy of Science Association 1974, 301-312.
Copyright © 1976 by D. Reidel Publishing Company, Dordrecht-Holland.
With kind permission from Springer Science+Business Media B.V..

if one adds to the theory-ladenness thesis the very plausible assumption that common-sense categories of observation are not identical with the categories of observation associated with scientific theories,[3] then one can pose the Kantian-like question, 'How is science education possible?' It is only with such cases of non-identical sets of observational categories that I shall be concerned in the remainder of the paper.

What I wish to do, then, is this: first, I shall examine Scheffler's two main moves with regard to the theory-ladenness of observation in order to show that whatever his success in saving objectivity in the abstract, he seems to leave the science student in an extremely precarious situation. Sketching the nature of this situation vis-à-vis the yet to be learned subject and vis-à-vis the teacher will, hopefully, illustrate the force behind the question, 'How is science education possible?' Finally, I shall suggest an answer to this question which utilizes metaphor as a key feature of science education.

I

Scheffler's defense of objectivity has two main prongs. First, he urges that in considering the theory-ladenness or conceptual nature of observation, we proceed by

> ... distinguishing categories from hypotheses, and contrasting the general ordering imposed by the former with the particular categorical assignments predicted by the latter; observation "determined by," "dependent on," or "filtered through" categories is thus quite conceivable as independent of any special hypothesis under test, expressible through reference to such categories. The general view thus advocated seems to me to preserve a tenable notion of the objectivity of observation, and to do so, moreover, without presupposing that the given is ineffable, uninfluenced by categorization, or reported by statements that are necessarily certain.[4]

Thus while different people may have different categorial schemes, these schemes *may* overlap, and, in any event, the *hypotheses* which describe the relations into which categorized experience will fall can be tested by seeing whether or not our experience does so order itself. Furthermore, note that Scheffler holds this view without holding that the given is ineffable, pure, or certain.

Now while the observational categories of science student and scientist or science teacher *may* overlap to some degree, it is highly unlikely that they will overlap as regards those categories peculiar to the scientific theory. Thus, even if observation in terms of a category schema provides an objective check on hypotheses for those who already share the category schema, the poor science student who has not yet acquired that schema remains in a quandary. How is he to learn the observational category schema when that schema depends on the theory which he does not know?

It might be suggested here that the way out is to teach the student the theory and then the category schema will follow. This suggestion leads me to the second prong of Scheffler's defense of objectivity, namely the defense of the objectivity of meaning. Scheffler's problem is this: the meaning of categorial terms is language dependent. Thus for the student to learn the theory, he must, in effect, already have learned the language in terms of which the theory is expressed. But this seems to be impossible. Furthermore, Schaller has already abjured the traditional way out of this difficulty, namely, that the student and teacher share the same basic observations and neutral observation language, and, hence, could build up the meanings of the new terms out of already shared meanings along with ostensive definition. In short, Schaller must account for the meaning independence of some categories and experimental laws from their role in specific theories without utilizing neutral observation.

Scheffler's solution to this problem lies in an appeal to the distinction between sense and reference. Differing senses of categories of observation and of experimental laws are indeed possible in different theories, although some synonymies may persist from theory to theory. Of even more importance, however, even in cases of varying senses of terms, reference can be the same from theory to theory.[5] Furthermore, different theorists can agree on the constancy of referential interpretation by application to specific cases. Surely both Aristotle and Galileo could point downward and agree that was the earth.

Yet this appeal to agreement of reference in specific cases is not without its problems, for Scheffler also acknowledges the possibility of a multiplicity of schemes of reference, thus allowing only a relative independence of observation from theory. In effect, the appeal to agreement of reference in special cases is open to change, reinterpretation, and charges of misinterpretation. In essence Scheffler accepts the Wittgensteinian attack on the ultimacy of ostensive definition. Thus while the logical possibility of scientific objectivity may have been salvaged by

Scheffler, the epistemological problem, and even more importantly, the pedagogical problem, seem to remain. How can we know when we have commonality of reference? How can we teach the student the language of reference peculiar to the science in question?

II

Scheffler seems aware of the general thrust of these problems as is revealed in his description of a new theory with its own unique scheme of reference emerging within a different, given, referential tradition. His description is worth quoting at length.

> A new theory arising within a given referential tradition cannot command initial consensus on presumably confirming cases of its own, but must prove itself against the background of prior judgments of particulars. It must acknowledge the indirect control of accumulated laws and theories encompassing already crystallized judgments of cases. Even if some such judgments are to be challenged from the start, the challenge needs to be expressed in a form that is intelligible for the received descriptive mode, and special motivation for the challenge must, of course, be adduced. The inventor of a new theory cannot, at the outset, motivate his new forms of discourse by simply saying "Look and see!" Now, it may turn out that this new theory, having won an initial place largely through indirect forms of argument against the background of acknowledged facts, eventually forces a revision of older judgments of cases and, what is more significant, perhaps, opens up new ranges of evidential description, thereafter developing consensus on relevant instances of its own. It remains true that, at the outset, its advantage needs to be shown in the context of judgments of cases already available, and in relation to the scheme by which such judgments are formulated.[6]

Now it seems to me that this situation is precisely that which obtains most of the time between science teacher and beginning science student. The student is the bearer of the 'received descriptive mode' and the teacher is attempting to make plausible a 'new theory' with new schemes of reference. Scheffler admits the teacher cannot simply say, 'Look and see!'

Rather he must couch his instruction in relation to the student's existing referential scheme. Of course, one would not wish to deny this rather novel way of expressing the hoary dictum that one must start with where the student is at, but the question remains, just how does that aid the student's understanding? Recall in this connection that building up new categories out of neutral, shared, independent, categories has been ruled out. What kind of 'indirect argument' for the new theory can be made to the student? How is science education possible? For that matter, how is any education which utilizes radically different schemes of reference possible?

III

There are a number of science educators who believe strongly in one version or another of the thesis of the theory-dependency of observation, and, furthermore, utilize this cluster of ideas in their recommendations for what science teachers should be taught to do. One of these is my colleague, Charles Weller, at Illinois.[7] Now although Weller's total view lends itself in places to the kind of subjectivist interpretation attacked by Scheffler, nevertheless, some of his suggestions as to how standard science education is possible seem to me to be fairly illuminating. He lists four recommendations.

(1) The teacher must assess the student's frame of reference as quickly as possible.

(2) If the student has no prior experience with the relevant phenomena, then direct experience with the phenomena should be provided.

(3) The student should be involved with the phenomena attempting to organize them in terms of the accepted model.

(4) The student should be communicating actively with others, at least some of whom already have a good working knowledge of the model in order to test out their tentative articulations.[8]

Now while these suggestions seem most plausible and several of them compatible with Scheffler's argument, they seem to me to lack a coherent unifying perspective. What I will do in the remainder of the paper, therefore, is to try to sketch such a perspective by using the notion of a metaphor as the key pedagogical device.[9] In brief, I want to say that science education, indeed any education involving differing schemes of reference, is possible because of the existence of metaphor.

In making this claim, I shall be using a fairly standard account of metaphor due to I. A. Richards and modified by Max Black.[10] I shall speak of the linguistic term being used metaphorically, as the 'vehicle.' That about which the metaphorical assertion is being made I shall call the 'topic,' whereas the 'ground' will be that which the topic and the ordinary literal referent of the vehicle have in common. The dissimilarity between the ordinary reference of the vehicle and the topic is called the 'tension.' On Max Black's view there are two types of metaphorical assertion which are relevant for my purposes.[11] *Comparative* metaphors are to be understood as basically analogies between topic and ordinary referent. *Interactive* metaphors, on the other hand, involve an interaction between the system of categories and beliefs clustering around the ordinary referent of the vehicle and the system of categories and beliefs clustering around the topic. Black's reason for also recognizing the interactive metaphor is that in some cases one simply cannot understand how a set of analogies as used in comparative metaphors can give us the insight which a metaphor often provides.

> It would be more illuminating in some of these cases to say that the metaphor creates the similarity than to say that it formulates some similarity antecedently existing.[12]

Thus, at least with interactive metaphors, there is a unique cognitive role played by the metaphor. Significantly, my colleague, Andrew Ortony, titles a recent paper of his, 'Why Metaphors Are Necessary and Not Just Nice.'

Note well that metaphorical assertions necessarily *show* us something about the topic rather than describing the topic literally. Metaphorical assertion is thus peculiarly suited to cases where we do not wish to or cannot literally describe the topic.[13] This seems to me to provide the key to bridging the gap between the student with his frame of reference and the to-be-learned frame of reference of the scientific theory. Aspects of the topic can be *shown* to the student rather than described, and this is crucial since by hypothesis the student is unfamiliar with the appropriate mode of description in the science. Note too, that the 'showing' in the use of metaphor is *not* simply ostensive definition. If it were, we would not need the metaphor to direct our attention to crucial aspects of the topic.

How does this work in detail? Well, the to-be-learned scheme of reference may overlap partially or not at all with the student's existing scheme of reference. If it overlaps partially, then the student can perform some of the tension-elimination for himself, i.e., he can see, to some extent what the dissimilarities between topic and literal referent are. To the extent that this is possible, the metaphor may be a comparative one. I shall concentrate, however, on the extreme case in which the two schemes of reference at least as regards the scientific categories are disjoint.

In this case the teacher introduces the key metaphor to the student. The vehicle must be part of the student's linguistic and cognitive scheme, hence the necessity for Weller's point that the student's frame of reference be assessed by the teacher. But, furthermore, since the student has, by hypothesis, no familiarity with the topic, he must, as Weller says, be put into direct contact with the phenomena, even though he may be familiar with the phenomena under a different categorization. In Weller's terms the student must be involved with trying to organize the phenomena by means of the model (the metaphor) he has been given. Initially this will involve his organizing the topic in accordance with the literal meaning and implications of the vehicle.

But since on Weller's view the student is in the presence of someone already familiar with the topic, the tension involved in these naive inferences can gradually be shown to the student and thus eliminated. Books could, in principle, also serve this weeding out or tension-elimination function.

Furthermore, in some cases, the environment itself may be sufficient to show the need for the appropriate tension-elimination as the student acts on his initial understanding of the topic wherein the vehicle is used literally in speaking of the topic. In other words, the ground is taken by the student to be complete prior to the weeding out by the environment of false predications and inferences. Logically, the research scientist at the frontiers of his field is in the same position as the student without a teacher but with an environment to help him understand the topic. The claim that nature is science's teacher may not be so metaphorical after all!

Let me now consider one possible objection. Black asserts that *both* comparative and interactive metaphorical assertion require simultaneous awareness of both topic and ordinary referent.[14] Yet I have claimed that the student's frame of reference need not overlap at all with the to-be-learned frame of reference. This would appear to be equivalent to denying student awareness of the topic. In an important sense, this logically possible case of total disjointedness of systems of reference probably almost never occurs in fact. The physics student at least sees the cathode-ray tube as a glass and metal object. Even the superstitious native sees his own photograph as magic of the gods. So in most cases there are at least some referential categories common to the student's system of reference and the new system of reference, and in this sense there is an awareness of both systems. So too will there be some tension resulting from the metaphor's invitation to apply the familiar categories to the new situation, along with a fairly vaguely apprehended ground. The teaching-learning situation then proceeds by making explicit more of the tension and progressively eliminating it thus making the ground more and more precise until, at some unspecifiable point, we would probably cease to count the metaphor as a metaphor. The term has a new meaning and science education has been seen to be possible. At the same time, however, it does seem possible that occasionally the systems of reference are totally disjoint, and it is here also that I am suggesting that metaphor plays an essential role.

As I have described the situation, however, there is something which serves a role similar to that of the interaction of two ideational systems in awareness. It is at this point that the influence of the world can make itself felt as an independent cause of objectivity. The teacher, or environment in the ultimate case, edits out those categorizations and inferences which are carried over whole with the vehicle and applied to the topic. One literally does not know that about which one enquires, but one can, at least, eliminate false guesses.[15] The interaction can be between idea and world in a strictly causal sense. The awareness that occurs may be no more than the realization that there is something wrong with organizing phenomena in the accustomed way. I think this kind of ideational-environment interaction satisfies the spirit if not the letter of Black's requirement. Objectivity needs no more of a foothold.

All this talk of the metaphorical extension of schemes of reference may have left the impression that I have been dealing with something extremely rare, revolutionary, and mysterious. This impression was almost unavoidable given that I have been dealing with the relatively self-contained theories and models of contemporary science. Learning such material is largely confined to fairly standard courses and, indeed, appears quite mysterious to many students—those who succeed as well as those who fail. Then too I have, by stipulation, been discussing the situation in which the scheme of reference of the student is radically disjoint from the scheme of reference of the science to be learned. However, it may be worthwhile to point out that the extension and change of schemes of reference by means of metaphor is simply an extreme case of a process of conceptual change and modification which goes on all the time in learning.[16]

Consider the child who has come to know about soft, fuzzy animals called dogs. One of them, called Socrates, is her pet, and there are several others in the neighborhood. Now this child, call her Ann, is taken to visit her Aunt Louise where she is fascinated by a tiny porcelain figure she can hold in her hands. 'That, too, is a dog,' says her father. Is the phrase 'china dog' a metaphor? Probably not in the ordinary sense. But then there's an awful lot of difference between Socrates and Aunt Louise's dog, too. Have Ann's conceptual scheme and referential categories been changed? Almost surely, but there is nothing rare, nor revolutionary, nor mysterious about it. My point is that learning a new categorial scheme in a science is very much like, although more extensive than, learning about china dogs.

I think that Kuhn has also recently made this same point regarding the possibility of learning a completely new categorial schema without any necessarily linguistic translation of the new into the old. In 'Second Thoughts on Paradigms,' Kuhn says,

> ... I suggest that an acquired ability to see resemblances between disparate problems plays in the sciences a significant part of the role usually attributed to correspondence rules.[17]

This suggestion occurs in the context of Kuhn's discussion of exemplars as an extremely important sense of 'paradigm.' Briefly, exemplars are shared exemplary 'problem solutions'—actually basic observational categories which are learned *without* necessarily learning criteria of application.

As Kuhn says

> ... I continue to insist that shared examples have essential cognitive functions prior to a specification of criteria with respect to which they are exemplary.[18]

Kuhn distinguishes the 'exemplar' sense of paradigm primarily for the historical reason that he could not find sufficient evidence of rule and criteria assimilation in the training of scientists to explain the similarity of judgments made by scientists in actual uses. Thus, Kuhn posits the exemplar as a non-rule-governed manner of accounting for similarity judgments.

My preceding discussion has made explicit another, logical, reason for requiring something like a learned exemplar. Barring a neutral observational base or the possibility of translating between referential schemes, only something like an exemplar *could* do the trick. But what may appear mysterious when spoken of in the language of exemplars becomes, I suggest, quite plausible in the language of metaphor—especially interactive metaphor. For metaphors have been used for a very long time to organize new fields into similarity sets prior to our ability to state formally the criteria with respect to which the similarity grouping is made.

The new element, seemingly forced on us by the theory dependency of observation, is the essential role of metaphor or exemplars. Metaphors have generally been viewed as merely heuristic devices to aid learning, and, indeed, they may have no more of a role than heuristic in other areas of science education. However, if my preceding characterization of the predicament of the science student is at all correct, the logical role of metaphor *must* be played if the student is to learn a radically new scheme of reference.

Let me conclude then by noting that Paul Feyerabend's fascinating account of Galileo's Tower Argument seems to illustrate beautifully all of my points.[19] The Tower Argument was advanced by proponents of a stationary earth. It consists essentially of arguing that a rotating earth would mean a stone dropped from a tower would have to land some distance from the base of the tower. Clearly this does not happen and so the thesis of a rotating earth must be false.

Feyerabend notes here that Galileo's problem is to replace one natural interpretation of motion, i.e., one theory-laden categorial system, with another. People must be brought to see the operative nature of only relative motion. And how does Galileo qua science teacher attack this problem? By introducing the metaphor of an artist drawing a picture in a boat during a long trip and placing the observer outside the boat and noting the relative motion of boat (as earth), paper (as tower) and pen (as stone). Galileo also uses other boat and carriage metaphors to teach the new system of natural interpretations or observational categories. Feyerabend admiringly refers to this process as 'psychological trickery' but I suspect he would not object too strenuously to my stronger claim that such 'psychological trickery' is essential for learning radically new systems of reference.

In summary, I have urged that the pedagogical question, 'How is science education possible?' remains as a legacy of the theory-dependency of observation even after objectivity has received its due. I have further urged that an interactive view of metaphor provides the key to answering this pedagogical question in that it allows one to bridge the gap between alternative systems of reference even when these are disjoint. In short, metaphor plays an essential cognitive role in learning. I have also expanded the notion of the interaction of ideational systems in metaphor to the interaction of ideational systems and environment and urged that this interaction is very similar to Kuhn's idea of learning non-rule governed perceptual similarity relationships by means of exemplars. Finally, I have hinted that this interaction may prove to be the material locus of the objectivity formally saved by Scheffler's analysis. Science education is possible with the aid of the humanistic tool of metaphor, and consideration of current philosophy of science seems to force such a conclusion on one.

University of Illinois

Notes

1   See for example, Thomas Kuhn, *The Structure of Scientific Revolutions*, 2nd ed. (Chicago University, Chicago, 1970); Paul Feyerabend, 'Against Method: Outline of an Anarchistic Theory of Knowledge', in M. Radner and S. Winokur (eds.), *Analyses of Theories and Methods of Physics and Psychology*, Minnesota Studies in the Philosophy of Science, Vol. IV (University of Minnesota, Minneapolis, 1970); N. R. Hanson, *Patterns of Discovery* (Cambridge University, Cambridge, 1958); Stephen Toulmin, *Human Understanding* (Princeton University, Princeton, 1972); and Michael Polanyi, *Personal Knowledge*, revised edition (University of Chicago, Chicago, 1962).

2   Israel Scheffler, *Science and Subjectivity* (Bobbs-Merrill, New York, 1967).

3   Both Scheffler and one of my co-symposiasts do seem to grant this very assumption. See Scheffler, *Ibid.*, e.g., pp. 40-41 and 64-65. See also Michael Martin, *Concepts of Science Education* (Scott-Foresman, Glenview, Ill., 1972), p. 127.

4   Scheffler, *op. cit.*, p. 43.

5   *Ibid.*, p. 62.

6   *Ibid.*, pp. 65-66.

7   Charles M. Weller, 'A Psycho-Epistemological Model for Teaching Science and Its Articulation with Classroom Activities'. A position paper delivered at the Association for the Education of Teachers in Science Meeting, Chicago, March, 1974.

8   *Ibid.*, pp. 26-27.

9   In the following, I am extremely indebted to my student, Felicity Haynes, and my colleague, Andrew Ortony, for having opened my eyes to the crucial importance of metaphor in learning situations where a new scheme of reference is being learned. See Andrew Ortony, 'Why Metaphors are Necessary and Not Just Nice', *Educational Theory* 25, Winter 1975, pp. 45-53.

10 See I. A. Richards, *The Philosophy of Rhetoric* (Oxford University, London, 1936) and Max Black, *Models and Metaphors* (Cornell University, Ithaca, New York, 1962).
11 *Ibid.*, pp. 25-47.
12 *Ibid.*, p. 37.
13 I am indebted for this point to Felicity Haynes. Ortony, *op. cit.* makes a similar point in speaking of the inexpressibility thesis of metaphors. Ortony urges that there literally are things which cannot in fact be expressed in a given language. For my purposes, I only need admit that some things are not now expressible literally in the language presently at the command of one speaker, namely, the student.
14 Black, *op. cit.*, p. 36, 46.
15 This is, in roughest outline, my own version of Popper's method of conjectures and refutations. See Karl Popper, *Conjectures and Refutations*, 2nd ed. (Basic Books, New York, 1965).
16 I am indebted for this point to my colleague, F. L. Will.
17 Thomas Kuhn, 'Second Thoughts on Paradigms', in F. Suppe (ed.), *The Structure of Scientific Theories,* (University of Illinois, Urbana, Illinois, 1974), p. 471.
18 *Ibid.*, p. 477.
19 Paul K. Feyerabend, 'Against Method: Outline of an Anarchistic Theory of Knowledge', in Michael Radner and Stephen Winokur (eds.), *Analyses of Theories and Methods of Physics and Psychology,* Minnesota Studies in the Philosophy of Science, Vol. IV (University of Minnesota, Minneapolis, 1970).

[1976]
A Rule by Any Other Name is a Control System

Lurking in the background, and often times the foreground of a whole bevy of psychological theories is an absolutely crucial family of notions. The ideas of rules, rule-following behavior, rule-guided behavior, rule-conforming behavior, rule-governed behavior, the application of rules, obeying a rule, tacit rules, and probably dozens of other variations are central to a great deal of psychology. And yet the concept of a rule is ill-understood at best.

What I shall do in this paper is, first, to illustrate, by no means exhaustively, a portion of the wide range of psychological theorizing in which 'rule' figures centrally. The cases I will discuss are psycholinguistics, social psychology, and artificial intelligence. Second, I shall argue on the one hand that there is an extremely important core to be extracted from all the uses of the concept of rules, from tacit rules to explicit rule application; but on the other hand this core cannot be accounted for in stimulus-response, behavioral, or causal terms—at least as these are ordinarily understood. Finally, I shall show that the negative feedback model of behavior as the control of perception suggested by William T. Powers provides just the right model to capture the central core notion of a rule.[1] What this implies is that the control system model of behavioral organization provides the standpoint from which we can unify and understand a great deal of otherwise very disparate psychological theorizing. And that effect, the ability to see unity in variability, is almost a hall-mark of the cybernetic approach.

Rules, rules and rules

It should take no great amount of argument to convince everyone of the central importance of the concept of rules for psycholinguistics. In 1970, George A. Miller wrote a fascinating essay entitled, "Four Philosophical

---

Reproduced with permission of the American Society for Cybernetics from *ASC Cybernetics Forum VIII*, Fall/Winter 1976, 103-114.

Problems of Psycholinguistics," with one of the problems being the nature of rules.[2] There seem to be two main problems for Miller with regard to rules. Although loathe to give up the theory of habit as a potential explanation for rule-governed behavior, Miller cites cases which "impress one intuitively with the huge gap between the simplest habits and the most complex systems of rules, yet the precise nature of this gap is difficult to characterize."[3] Second, he cites the familiar assertion by many linguists that "when a person knows a language, in the sense that his own utterances exhibit these observed regularities, there must be some sense in which he knows the rules of the language."[4] The problem, well-known, of course, is to say just what that sense is in which the person knows the rules of the language, for it is certainly *not* the case that the person can formulate the rules explicitly. Or, as Chomsky puts it,

> The person who has acquired knowledge of a language has internalized a system of rules that relate sound and meaning in a particular way. The linguist constructing a grammar of a language is in effect proposing a hypothesis concerning this internalized system. The linguist's hypothesis, if presented with sufficient explicitness and precision, will have certain empirical consequences with regard to the form of utterances and their interpretations by the native speaker. Evidently, knowledge of language—the internalized system of rules—is only one of the many factors that determine how an utterance will be used or understood in a particular situation.[5]

Here knowledge of a language is explicitly identified with an internalized system of rules—a philosophical grammar as it later turns out to be. This grammar is not a set of prescriptions about how people ought to speak, but an explanatory theory (more precisely, part of a theory) of how they can comprehend and speak a natural language. It is a system of *rules* because as Chomsky and others have argued so forcefully, a system of associations is simply inadequate to account for a native speaker's ability to produce and comprehend novel utterances.[6]

My second illustration comes from social psychology.[7] In this area the idea of rule-following behavior seems almost to constitute the meaning of "the social." A familiar example will help. The writing of a check to pay a bill is sometimes cited as a paradigm case in which the rules and institutions surrounding the transaction provide its (social) meaning. In the absence of the rules of banking, writing numbers and one's name on a piece of paper simply do not have the significance of writing a check. Indeed, even so simple a case as marks on a paper constituting a *signature* requires a whole host of social rules as an explanatory context. In short, many, if not all, pieces of behavior are constituted as (social) actions in virtue of their conformity to certain social rules. And these rules provide the "appropriate" level of description for what is going on.

Beyond such constitutive rules, however, there are also strategic rules, rules of etiquette, and so on governing our social transactions.

Sometimes bills fall due a day or so before paychecks are deposited. People sometimes write checks "on the float." That is, they write the check in time to meet the deadline for paying the bill knowing that by the time the check clears the account, there will be funds to cover it. A useful strategy for avoiding finance charges. Further examples of etiquette and strategy rules could be given. "Write the stub of the check first." 'One ought to begin the written description of the amount (as opposed to the numerical description) at the far left of the appropriate line." And so on.

Such social rules, unlike linguistic rules, are often taught explicitly. However, it is not at all obvious that once learned they are consciously *applied* to new situations. How, then, do such rules control and account for social behavior? Could we, perhaps, reduce the rules to some complex "habit theory"? Or is the level of description and explanation provided by social rules *sui generis?*

My last example of the centrality of the concept of rule comes from work in cognitive psychology, especially work which attempts to simulate human cognitive processes on a computer.[8] In such cases the rules being followed are the rules of the computer program. To the extent that the program actually simulates human cognitive processes, these program rules are presumed to have their counterparts in some sense in the human mind. Once again, the concept of a rule, and following a rule, is crucial. To cite just one example from Anderson and Bower's HAM, the input to HAM's "mental" system is in the nature of tree diagrams generated by linguistic or perceptual parsers which follow *rules* for transforming ordinary stimulus elements into the trees.

In the case of computer simulations one speaks fairly comfortably about an explicit following of rules. In some sense, the computer applies the rules in the program in its operation—much as the beginning logic student applies the rules of proof in constructing derivations. But even here one has the problem of identifying the "appropriate" situation in which to apply the rule. This problem is masked to a very large extent in computer simulations by the fact that the inputs to the program are guaranteed to be perceptually appropriate.

Human perceptual systems have a much harder time of it. For part of the learning task for humans is to come to recognize when the circumstances are the same" so that one can apply the appropriate rule.[9] As Green puts it, To learn a principle [rule] is not, therefore, simply to develop a disposition to do the same thing in similar circumstances, but to learn what counts as "doing the same thing" or what constitutes "similar circumstances."[10] How does one differentiate a situation which calls for tact from one which calls for forthrightness?

Without question, my list of examples in which rules occupy a central position in psychology is incomplete. Equally I am sure I have not illustrated all the numerous kinds and levels of rules which might be said to exist.[11] With regard to the former incompleteness I trust that other examples can be generated from my audience's experience and knowledge. With regard to the latter defect, I think I have represented at least some of the major kinds of rules and some of the distinctions among them. It is to the elucidation of these distinctions that I now turn.

A rule is a rule is a rule is a . . . .

Perhaps *the* central distinction often drawn between different kinds of rules is that between descriptive and prescriptive rules—between rules which describe or explain linguistic or social behavior as it is empirically observed and rules which prescribe correct grammar or proper social behavior. And, of course, the distinction is a perfectly valid one. Psychology is, and ought to be, concerned with the descriptive sense of rules. At the same time, however, I believe that this distinction has also been one of the most pernicious influences hindering a proper recognition of the unique function of rules as an explanatory concept in a purely descriptive sense. The problem is that once the prescriptive variety of rules has been noted, it becomes well-nigh impossible to see that descriptive rules are still normative in an absolutely fundamental sense. But, if they are normative, they must be prescriptive, right? Wrong! Once the normative nature of descriptive rules is overlooked, the temptation for psychologists to assimilate descriptive rules to generalization and to habit theories in particular becomes virtually irresistible.

Let me show the operation of this essentially norm-regarding feature of even descriptive rules in my examples. Given the linguists' program, the production of a non-grammatical or anomalous piece of verbal behavior does not count against some generalization as to what will or will not happen, but rather is not counted as a piece of language. It doesn't meet the minimal conditions for being a meaningful utterance and thus the rules *legislate* what counts as *appropriate* linguistic behavior.

This legislation is not on the level of linguistic etiquette or even linguistic strategy—that kind of legislation might be construed as a prescriptive rule. Rather the legislation is on the *constitutive* level of what counts as meaningful discourse.

With respect to social systems, what constitutes writing a check is defined by reference to the rules, the norms, for classifying any piece of behavior as that of writing a check. Once again at the descriptive level, behavior must be *appropriate* to be judged as the writing of a check.

Finally, even in the case of the computer simulation, the rules of the program as opposed to the physics of the hardware, are norm-regarding in the appropriate sense. The input to the linguistic parser must be judged to be signal as opposed to noise, to be a linguistic string as opposed to something else. As I have mentioned, this requirement is not obvious since computer programs generally ignore the perception problem by designing the inputs to be automatically perceptually significant. Yet significant they must be.

Once one begins to see the sense in which even descriptive rules are norm-regarding, in being constitutive of certain kinds of meaningful behavior, one is sometimes tempted to utilize the explicit-implicit or the related conscious-unconscious distinction with a fairly heavy hand. That is, *rules* turn out to be appropriate modes of description only when they can be made explicit and/or consciously learned. Typical models here include the constitutive rules of chess and the explicit rules of logical deduction. The movement of a chess piece which doesn't accord explicitly with the rules just doesn't count as a move in chess. A step in a proof which does not formally follow the rules, just isn't a part of the proof.

Of course, rules can be and often are made explicit. But it is surely a mistake to suppose that they must be. People reasoned correctly and made judgments about the correctness of their reasoning long before Aristotle began to make the rules of logic explicit, and even now that they have been made explicit, it's not at all clear that a course in formal logic will help one reason any better. The point is that judgments of good and bad reasoning can be made without appeal to explicit rules. But normative judgments are unintelligible without presupposing rules or principles. A piece of reasoning is good or bad *because* it does or does not come up to the standard, the criterion, of good reasoning. But in such cases where the norm is implicit and even not consciously held, there is a criterion for the operation of rules. One must look at the attitude taken toward situations which are violations of the norm. Such situations must in some way be judged as "incorrect"—as violations of the norm.

One could, nevertheless, insist that rules are to be rigid, formal, explicit, and consciously followed, and that the norm-regarding behavior which is not of this nature be called something else. If one is of such a mind, I can only say, "Very well, you can have the word, 'rule', if you wish, but I am calling attention here to some very crucial and important similarities between your narrow notion of rule-governed behavior and a broader notion of norm-regarding behavior." At the same time, I feel that the similarities are so important that the narrow notion of rule-governed behavior deserves to be broadened to encompass all cases of norm-regarding behavior.[12]

One last indirect argument for the broadened conception. Explicit rules can be and sometimes are changed. Why? Well, think of a proposed rule change in chess. Why might we approve or not? The arguments would appeal to the general purposes for games like chess and the extent to which the proposed change might help or hinder such purposes. One would also cite the consequences for playing the game and the revisions that would be entailed.

I have been told, for example, that at one point in the history of chess, the Queen's versatility of movement was increased precisely because the game up to that time had become too routinized—much like a complicated tic-tac-toe. In short one argues over the appropriateness of such changes. Appropriate to what? The *implicit* standards, purposes, goals, norms, we do not consciously hold, but are found in historical practice for games like chess. So judgments as to the appropriateness of *changing* explicit rules are made by reference to implicit historical norms of practice. Such judgments are a species of norm-regarding behavior. And I want to call attention to the continuity here between the explicit and the implicit norms.

Norm-regarding behavior involves judgments—judgments that *this* behavior is appropriate or falls under the norm. And in such judgment it seems possible to distinguish such norm-regarding-behavior from mere habit, even if the norm-regarding behavior is itself not conscious or explicit. The concert pianist knows when a wrong note has been hit even though no explicit conscious judgments were being made that "Now *this* note is appropriate to the score." The pianist just plays, concentrating more on technique and especially interpretation. In the case of mere habits, one does not judge the appropriateness of behavior, rather the same circumstances simply call forth the same behavior. There is no obvious way in which there can be mistakes in habitual behavior. One can acquire bad habits, or unintended habits,—but the notion of their being a mistake in the operation of the habits seems queer. It is what it is.

It is actually misleading, however, to speak of "judgments" of appropriateness of norm-regarding behavior; for such language is heavily biased toward conscious activity and the judgments need not be conscious at all. Perhaps a more adequate formulation would be in terms of seeing the situation as one in which the given norm-regarding behavior is appropriate; I recognize this as a situation in which the writing of a check is called for. In short I experience or perceive the situation in the descriptive terms provided by the norm or rule. And it is because of my perception of the situation as appropriate to the given norm that I behave as I do.

But there is another crucially important feature of norm-regarding behavior which must be noted here. Although my perception of a situation as appropriate for behavior of a certain variety in some sense causes that behavior, nevertheless, the behavior need not always live up to the norm. I may misspeak. I may enter inconsistent amounts on my check. The pianist may strike the wrong note, my proof strategy may fail. My behavior may not live up to the norm, but unless it is grossly inadequate, it may still be appropriately described and thought of as norm-regarding. Conversely, there are actions which bring about certain unintended end-states. As I argued above, the test for whether the action was aimed at the end state would be whether impediments and disturbances to reaching the goal were treated as mistakes.

Such a situation almost never obtains when one is speaking of purely descriptive generalizations. For example, it would certainly be possible that there has never been and never will be a completely error-free performance of some complex piece of music. If that were so, and we tried as good empiricists, simply to describe all the performances as they actually occurred, we would at best get some rather strange looking statistical approximation to the actual score. Yet clearly such a generalization does *not* capture what any performer was doing. The performer was playing the piece, even if there were mistakes. The statistical generalization level of description is simply all wrong. The performer was engaged in norm-regarding behavior *even though* the norm was not perfectly fulfilled. I was writing a check even though I made a mistake.

Summarizing, although falling short of an analysis of 'rule,' the following, not necessarily independent, conditions seem to be criteria of something's being a rule:

1. *The Normative Condition* – Although rules can be divided into prescriptive and descriptive, descriptive rules presuppose norms just as do prescriptive rules. Behavior in accordance with descriptive rules is norm-regarding; it requires a judgment as to the appropriateness of the norm.
2. *The Self-Correcting Condition* – Norm-regarding behavior can be in accordance with implicit or unconscious rules if one stands ready to correct deviations from the norm.
3. *The Perceptual Condition* – In norm-regarding behavior, it is because one perceives (judges) a situation as structured, or constituted by the operative rule that one behaves as one does. Again, the rule need not be consciously applied.
4. *The Non-Success Condition* – One can follow a rule without succeeding in attaining the norm implicit in the rule.

. . . But a rule isn't a generalization

Implicit in the preceding discussion of the four criteria of a rule is the claim that a rule cannot be reduced to a mere descriptive (causal or statistical) generalization. Another way of putting this claim is that no mere habit theory of rules, no matter how complex, will ever capture the nature of rule-following behavior. In this section I want to argue this point a little more fully.

Powers nicely captures the thrust of the normative condition in his use of Brunswik's lens model of stimulus-response generalizations.[13] (See Fig. 1)

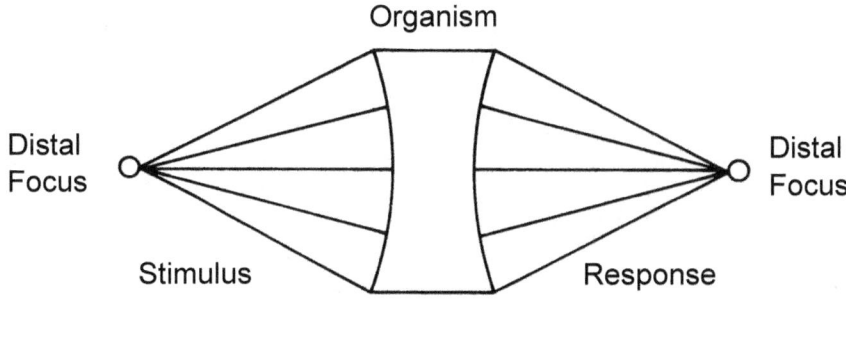

Figure 1.

The "stimulus" object represented by one distal focus can affect the sensors of the organism on different occasions by very different routes, represented by the bundle of "rays" leading to the "lens" or organism. Likewise the organism's behavior can achieve the same response—the other distal focus—by quite different specific behaviors.

Another way of expressing the insights in Brunswik's lens model is in terms of the problems of the definition of the stimulus and the response.[14]

This picture neatly describes the situation in most stimulus-response theorizing. But the lens metaphor can misleadingly suggest that with enough work the generalization underlying the connection of the distal stimulus to the distal response can be found, much as the laws of light explain the focusing properties of lenses. The difference in the behavioral case, however, is that the rays in the stimulus (or response) bundle are wildly different from each other in any physical sense. A rat can press a lever with any number of different muscle arrangements, even by sitting on it or dropping an object on it. Thus the power of the organism to "focus" its detailed specific behavior onto an action which turns out to be the same action each time or to focus its detailed stimulation onto the same stimulus is the power of perceiving or *judging* the action and the situation to be appropriate to the norm involved. The focusing metaphor is to be cashed, it seems, in terms of the appeal to norms defining very different detailed stimulations or responses as "the same."

I cannot repeat too often that "sameness" here is determined by the ways the organism perceptually represents and judges the situation in terms of the norm and *not* by any organism-independent properties of the so-called "real" situation. On the level of generalization, one could determine the minimum force needed to trip a lever, and quite independently of any normative judgments, the lever will be tripped by that force. On the other hand it is only by reference to the norm of what counts as lever-tripping that one could judge that the energy released by my voice in asking a research assistant to trip the lever could count as lever-tripping. It is the norm which focuses very- different physical generalities into a stimulus or response.

Considerations such as these lead to an examination of the perceptual condition. In rule-following behavior it is not just that the norm determines or constitutes what will count as the same stimulus or behavior, but also that in some sense it is *because* of the perception of the situation as appropriate to the norm that one does what one does. In some sense the judgment or perception of the situation as appropriate to the norm causes the action. Or, in common sense terminology, the organism behaves as it does in order to reach a goal, which is defined by the norm. But, of course, this introduces teleology into rule-following behavior with a vengeance, and modern psychology has spent a good portion of its history in trying to do away with teleology.

One of the main reasons for objecting to teleology has been that psychologists committed to straight-line causal models have been at a loss, conceptually, to explain how a non-existent goal can cause present behavior directed toward that goal. Causation just doesn't work that way and in the absence of an alternative model, the most rational course for psychology has seemed to be to deny the efficacy of goals. The alternative, for even the more cognitively oriented, has been to move to explanations in terms of a presently occurring desire for the goal along with beliefs about means to attain it. Such presently occurring beliefs and desires could, conceptually, serve as causes. The difficulty with this move is that it seems to be faced with the insurmountable problem of needing to postulate an indefinite number of beliefs and desires to account for all of the nuances of behavior and adaptive changes of behavior that occur in different situations leading to the same goal. In terms of the lens model, the "rays" are indefinite in number, and are rays of the *same* bundle solely in virtue of their leading to the same goal. The goal in mind *does* determine what "rays" could possibly be a part of the bundle and that still seems to be the wrong direction of influence for a causal generalization account of the matter to, handle. Norm-regarding behavior, on the other hand, seems, in virtue of the norm or goal, to be determined by judgments as to the appropriateness of the given situation. And I shall argue later that Powers' control system model provides the needed alternative for physically realizing a system capable of producing norm-regarding behavior in just those situations perceived to be appropriate *because* they are perceived as appropriate.[15]

The self-correcting criterion of rule-following behavior also seems at odds with a habit or even a more complex generalization account. Habits seem to be relatively narrow-tracked dispositions and are often contrasted with the adaptive nature of principled or rule-following behavior (at least in the broadened sense of rule-following that I am here urging). If one's behavior is under the influence of a habit, then one engages in it whether appropriate or not; whereas, the self-correcting nature of rule-following behavior implies a sensitivity and responsiveness to changing situations. To cite a well-worn example, a parrot may acquire the habit of saying, "Hello," in response to a certain stimulation, but there seems a world of difference between the parrot's automatic mechanical response and a person who understands the norms of greeting and says, "Hello" in response to those norms.

It may be that very complex habits or generalizations could be discovered which might account for the apparently indefinite plasticity and adaptiveness of norm-regarding behavior. However, it is clear no examples of such theories currently exist. It is at least plausible to suppose that one of the reasons for the continued insistence on a habit-type explanation for self-correcting behavior has been the absence of a non-mystical alternative model. It is that defect that I hope to help remedy with this paper.

There is, however, a deeper reason why no account of self-correcting behavior in terms of causal generalizations is likely to be successful. Self-correcting behavior presupposes a standard or norm of correctness. But, as already noted in discussing the perceptual condition, the perception or judgment of the situation as appropriate for the operation of the norm is causally efficacious in producing the "correcting" behavior. The purpose of the correcting behavior in turn is to correct the deviation from the norm. The norm operates in its own realization in the situation. Ordinary causal generalizations do *not* seem to operate in their own production or realization.

This point helps one to understand why the "rain drop analogy" as a way of discounting norm-regarding behavior is not convincing. According to the rain-drop analogy if one were to observe raindrops sliding down a glass, one might well be tempted to think of them behaving purposefully. They all "want to get to the bottom" and they get there in quite unpredictable ways. They even "correct" their behavior and take a different path if some obstacle is placed in their way. But clearly it would be absurd to impute purpose to the raindrops. The problem with the analogy is that it is gravity and *not* the putative purpose, namely, "the bottom of the window" which operates to produce the common end result. The situation described in the generalization is not operative in its own production. Yet in norm-regarding behavior, the norm is operative in its own production. And attempts to explain away that operation seem precisely to rob the description of the situation of its norm-regarding character. Recall once again the generalization description of the performance of a piece of music versus the description in terms of the purposeful attempts to play the piece. Of course, it is logically possible that behavioral science will yet discover the analogue to gravity in the raindrop analogy. But nothing on the horizon seems remotely to point to such a possibility.

Finally, consider the non-success condition. Norm-regarding behavior is still classifiable as such even if the norm is not attained. There is a serious conceptual problem here for those who would reduce norm-regarding behavior to generalizations. For when situations do not correspond to hypothesized generalizations, that counts as evidence *against* the generalization. Of course, a few anomalies can be tolerated, but not too many. On the other hand, behavior which is norm-regarding can fail to create the norm and not count at all against describing the behavior in terms of the norm. Indeed the norm may never be reached and we would still not refute the norm-regarding description. Again think of the complex piece of music.

There are some limits here. The behavior usually and over a period of time has to come close to the norm and it must always be corrective in the direction of the norm to be described as norm-regarding, but it need not get there. With generalizations such a situation would necessarily count against the generalization. I simply do not see how a generalization theory can handle this difficulty.[16]

What's a control system?

If causal generalizations give little promise of explaining norm-regarding or rule-following behavior, what account can we give? Obviously, one in terms of control systems. But what's a control system? For a detailed description of control systems, I refer the reader to the already noted book by William T. Powers, *Behavior: The Control of Perception.*[17] Here I will use an earlier more generalized diagram and explanation by Powers of the basic control system unit of behavioral organization.[18]

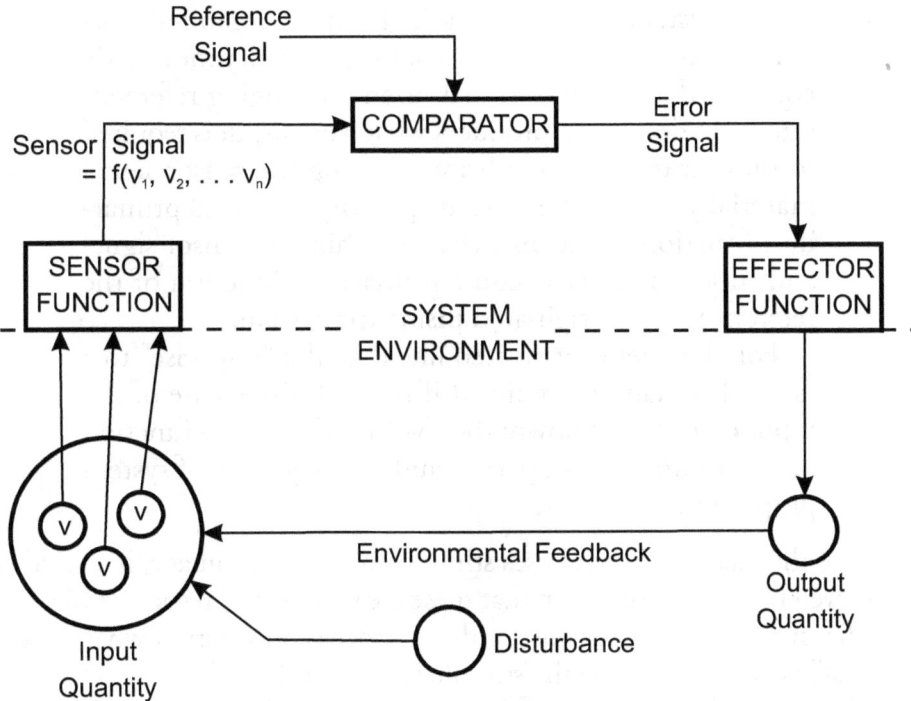

Figure 2: Basic control-system unit of behavioral organization. The sensor function creates an ongoing relationship between some set of environmental physical variables (v's) and a sensor signal inside the system, an internal analogue of some external state of affairs. The sensor signal is compared with (subtracted from, in the simplest case) a reference signal of unspecified origin. The discrepancy in the form of an error signal activates the effector

function (e.g., a muscle, limb, or subsystem) which in turn produces observable effects in the environment, the output quantity. This quantity is a "response" measure.

The environment provides a feedback link from the output quantity to the input quantity, the set of "v's" monitored by the sensor function. The input quantity is also subject, in general, to effects independent of the system's outputs; these are shown as a disturbance, also linked to the input quantity by environmental properties. The disturbance corresponds to "stimulus."

The system, above the dashed line, is organized normally so as to maintain the sensor signal at all times nearly equal to the reference signal, even a changing reference signal. In doing so it produces whatever output is required to prevent disturbances from affecting the sensor signal materially. Thus the output quantity becomes primarily a function of the disturbance, while the sensor signal and input quantity become primarily a function of the reference signal originated inside the system.

For all systems organized like this, the "response" to a "stimulus" can be predicted if the stabilized state of the input quantity is known; the "S-R Law" is then a function of environmental properties and scarcely at all of system properties.

Despite the fact that Powers uses stimulus-response language, it should be clear even from his description that these are stimuli and responses of a very peculiar nature. The "stimulus" or disturbance is only part of what has traditionally been taken to be the stimulus in classical psychology. The other part is supplied by the effects of the organism behaving. Indeed, this is one of the central features of a feedback system—it reacts to its own effects. In fact its effects are "designed" to keep the input quantity as close to the reference level as possible via the action of the effects through the environment on the input quantity. Variability of detailed output is seen in a unified way as keeping the input close to the reference level. I will elaborate on this point below, but for now the important thing is to note that this is no ordinary S-R mechanism. Indeed, the explanatory force, as we shall see, goes through the input side of the loop, *not* the output side. Paradoxical as it sounds, what feedback systems *do* is control perceptions—not behavior.

Furthermore, investigations into "black box" feedback systems to find out empirically what quantity is being controlled are clearly possible. Thus the internal structure postulated by the theory does have empirical implications and can be investigated empirically. Of course, the connection with the world is nothing like the naively direct one enjoined by a behaviorist methodology of operationally defining all internal structural concepts. Thus, negative feedback neither relies on mystical purposes nor is it unconnected with the world. It satisfies the requirement for having testable consequences without putting those consequences into the operational definition straight-jacket.

The way it does this is in principle very simple. If one suspects a negative feedback system is in operation, one then hypothesizes a controlled quantity for the system. Note that this "discovery" step depends on intuition and professional hunches, in this case, no more than does the comparable step of suggesting fruitful empirical operations for the behaviorist. Indeed, it is probably because of the close connection between controlled quantities and motives, goals, and purposes in ordinary-language talk about action that common sense provides as many fruitful hypotheses as it does. In any event, once a controlled quantity is hypothesized, the experimental procedure is this: introduce a disturbance near the sensor (it has to be the right order of magnitude so it neither escapes detection nor overwhelms the system) and see if the output opposes the disturbance. If it does, that quantity probably is being controlled. If there is no opposition to the disturbance, the hypothesized quantity is probably not under control. Utilizing the model one can even predict appropriate magnitudes.

Finally, note well that the line of control of a feedback system runs through the perceptual or input side of the model. There are no reference signals delicately controlling the detailed outputs. Indeed the reference signal can stay the same and the detailed outputs can vary considerably precisely to counteract the effects of disturbances on the controlled quantity. Feedback controls *sensed* quantities, *not* outputs. It affects outputs to be sure, but it does not control them. This term 'control' is a technical notion which refers to the operation of the feedback system to maintain the sensed signal near the reference signal no matter what the disturbance. The perceived quantity will be brought to match the reference signal by means of a wide range of outputs.

Powers traces in some detail the application of this model to the fairly complicated action of tracking a moving spot on a screen (e.g. an enemy plane on radar) by means of a stick controlling the position of a pointer.[19] However, my point is a conceptual one. The abstract model just described meets perfectly the four conditions of rule-following or norm-regarding behavior elucidated above. Control system theory thus promises a real alternative to habit or generalization accounts of rule-following behavior. It is to that claim that I now turn.

A rule by the name of a control system actually smells sweeter

Clearly, the normative condition of rules and rule-following behavior is satisfied by the control system model. The reference signal provides just the appropriate model for a norm. For the system acts to change the environment so that the sensor or perceptual signal matches the reference signal. "I need to write a check to pay my monthly utility bill," says the reference signal. The system outputs operate with materials in the environment, paper and pens, until the sensor signals indicate a check has been written. The behavior controls the perception to make it match the reference signal.

Moreover, the self-correcting condition of rule-following behavior is likewise clearly met by control systems. Since what they do is to maintain the sensed environment in the condition specified by the reference signal, they act to counteract any external disturbance which would tend to deflect the controlled quantity from its reference level. The situation of "writing a check" could be disturbed by any number of things which, if they registered as sensor signals, would create an error when compared to the reference signal and lead to corrective behavior. Such disturbances could be at any number of different levels. I might transpose some figures, misspell the name of the payee, remember I have insufficient funds in the bank, and so on. In each case, if sensed, such disturbances would lead to typically rule-like or norm-regarding self-corrective behavior.

Along similar lines the non-success condition can be seen to be easily accounted for by the control system model. Occasionally disturbances may occur which overwhelm the system's effective range of control. If the system tends, nevertheless, to counteract such disturbances, the proper level of description of what is happening is still that given by the control-system model. The musician who makes an error usually notices it and tries to correct it next time. Even if an error-free performance is never achieved, the musician is still properly described as "playing the piece." Or again, I may not have enough money to deposit to cover the check I want to write. If I try to borrow it or earn more, then I am clearly still behaving within the norms of check-writing, even if I never actually write that particular check.

The most exciting thing, however, about the control system model is the way in which it satisfies the perceptual condition. In the first place, the error signal, the difference between the reference signal and the sensor signal, is what drives the output. Thus the desired pattern, the reference signal, is operative in its own production—in the situation as actually sensed. For the system operates to reduce error—the difference between the desired pattern and the sensed pattern. Here is a physically realizable model that captures the essence of rule-following in which the rule *as* a rule operates in producing rule-conforming behavior.

But there is another important feature of the control system model related to the perceptual condition. The line of control runs through the perceptual side of the model. The model does not delicately apply its outputs to indefinitely varying situations carefully varying the output or behavior to match the situation. Instead it controls its inputs or perceptions. It operates on the environment, changing the environment until it produces a perceptual signal which matches as closely as possible the reference level. This feature accounts for the indefinite range of adaptability of rule-following behavior—as opposed to the straight-line operation of causal models. When an error is sensed, the system operates in *any* environment, using any means at its disposal to remove the disturbance and bring the perceptual signal in line with the reference signal.

Recall the classical behaviorist objection to purposes, goals, or rules as explanations of human behavior. A purpose or rule not yet in existence did not seem able to cause behavior leading to it. Recall also the classical move to meet the objection. The goal exists in present intentions and can be causal. This reply is fine as far as it goes, but as has been argued repeatedly, it does not go far enough. For the initial situation must still be perceived in terms of the intention. Is the situation appropriate for the operation of the rule? The goal or rule conceived as intention must still operate "backwards," in the sense that it at least partially structures the organism's perception of the existing situation. On a generalization view, there would be no way of knowing in detail what features of a novel situation would need changing to lead to the desired goal. But as soon as one's model provides for the control of perceptions rather than behavior, the necessity for detailed knowledge of how to change a current situation into a desired one disappears. The system just acts, and if at all well adapted to the general environmental conditions in which it finds itself, will produce situations which give rise to perceptions closer to the goal or reference signal.

Thus, returning to the example of psycho-linguistics, one can now see in some detail why the transformationalists' critique of associationist language learning and use is so powerful. The transformationalists urge that an associationist account could not possibly explain the production and comprehension of an indefinite number of novel utterances. And they offer an account in terms of transformational rules instead. But adaptive rule-governed utterances in novel situations are formally identical with the indefinite number of ways an output can remove a disturbance from a control system. If a certain deep structure is to be realized, the system will operate in (almost) any linguistic environment until it perceives that deep structure realized in the concrete situation. Neither associationism in linguistic behavior nor causal generalizations in rule-following behavior seem capable of accounting for the adaptive novelty we seem to find. Yet the control system model of behavior with its feature of the control of perceptions shows how one would expect such adaptive novelty as a matter of course.

With respect to the social psychology example, one can easily see on the control system model how social rules of meaningfulness can constitute certain situations. The social rules serve as reference signals and we learn them as well as ways of transforming physical situations into sensor signals which represent these social norms. Again, because the line of control goes through the perceptual side of the model, situations are "sensed as" being of certain kinds. This situation is a check-writing one and it is perceived as and treated as such. Control systems can, with appropriate input and output functions, control such non-tangible items as social meanings. Social meanings are not at all mysterious with the control system model.

Finally, as a number of workers in artificial intelligence have begun to realize, the perceptual component of human intelligence is the one which has been most neglected in work to date. Computer simulation has been most successful in those areas, even though of an abstract, problem-solving kind, in which the perceptual component is highly explicit. Proving logic theorems is an excellent example. The explicit rules of well-formedness and legitimate inference leave little room for perceptual ambiguity by either person or machine. The perceptual component is there; it is just non-problematic in much current work. What the control system model does is to point to the structural features which must be simulated if computer programs are to advance beyond theorem proving capabilities. Perceptual components must be built, and they must be built, not only with atomistic, bottom-up features analyzers, but also with top-down gestaltist control systems. For the control system model exhibits on its face the gestaltist insight that the whole determines the significance of its parts, and even what will count as parts.

Rule analyses continue to crop up in psychology. They do so at just those places where associationist, causal generalizations seem least successful. Behavioral analysis in terms of rules seems to require conditions incompatible with associationist, habit, or generalization models. But a control system model almost transparently meets the conditions for behavioral analysis in terms of rules. And it does so in a non-mysterious, physically realizable way. Psychology could well profit from a deep exploration of the control system model.

Notes

1. William T. Powers, *Behavior: The Control of Perception* (Chicago: Aldine, 1973).
2. George A. Miller, "Four Philosophical Problems of Psycholinguistics," *Philosophy of Science,* v. 37, June, 1970, 193-199.
3. Ibid., p. 190.
4. Ibid., p. 191.
5. Noam Chomsky, *Language and Mind* (New York: Harcourt, Brace and World, 1969), p. 23.
6. See, for example, Noam Chomsky, "Review of B.F. Skinner, *Verbal Behavior,*" *Language,* v. 35, 1959, 26-58.
7. See, for example, E. Goffman, *Relations in Public* (London: Allen Lane, The Penguin Press, 1971); C. Levi-Strauss, *The Savage Mind* (London: Weidenfeld and Nicholson, 1966) and R. Harre, "'Rule' as a Scientific Concept," in Theodore Mischel (ed.), *Understanding Other Persons* Totowa, New Jersey: Rowan and Littlefield, 1974), for a discussion of the social significance of rules.
8. A good recent example here is John R. Anderson and Gordon H. Bower, *Human Associative Memory* (Washington, D.C.: V.H. Winston & Sons, 1973).
9. I have argued the importance of perceptual learning in more detail in my, "The Believing in Seeing," in Lindley J. Stiles (ed.), *Theories for Teaching* (New York: Dodd, Mead, 1974).
10. Thomas F. Green, "Teaching, Acting, and Behaving," in B. Paul Komisar and C.B.J. MacMillan (eds.), *Psychological Concepts in Education* (Chicago: Rand McNally, 1967). The point is treated at length in L. Wittgenstein, *Philosophical Investigations,* 3rd edition (New York: MacMillan, 1968).
11. See, for example, the typology of rules generated by Stephen Toulmin in "Rules and Their Relevance for Understanding Human Behavior," in Theodore Mischel (ed.), *Understanding Other Persons* (Totowa, N.J.: Rowan and Littlefield, 1974).

12. I am, of course in good company in urging an expansion of the concept of a rule. Michael Polanyi in *The Tacit Dimension* (Garden City, New York: Doubleday, 1966) argues forcefully for the notion of tacit rules. I am convinced that from a very different angle, Polanyi's notions of tacit rules are getting at very similar kinds of ideas as I am in this paper. To argue that contention in any detail would take me too far afield. Equally good, but not as well-known, company arguing the same point includes Felicity Haynes, *Reason and Insight in Learning,* unpublished Ph.D. dissertation, University of Illinois, 1976.
13. Powers, op. *cit.,* p. 6.
14. See, for example, Donald T. Campbell, "Pattern Matching as an Essential in Distal Knowing,' in Kenneth B. Hammond (ed.), *The Psychology of Egon Brunswik* (New York: Holt, Rinehart and Winston, 1966), p. 81-103.
15. I have argued this point at more length in my "Action, Perception, and Education," *Educational Theory,* 24, 1974, 33-45.
16. I have argued this point in some formal detail in my "Action, Perception and Education," *ibid.*
17. Powers, *op. cit.*
18. William T. Powers, "Feedback: Beyond Behaviorism," *Science,* 179, 26 (January, 1973) 351-356.
19. Powers, *Behavior: The Control of Perception, op. cit.,* ch. 5, 6.

[1976]
Evolutionary Rationality: Or Can Learning Theory Survive in the Jungle of Conceptual Change? [1]

> A man demonstrates his rationality, not by a commitment to fixed ideas, stereotyped procedures, or immutable concepts, but by the manner in which, and the occasions on which, he changes those ideas, procedures, and concepts.
>
> Stephen Toulmin, *Human Understanding*

This paper is about conceptual change. In particular this paper is about the largely ignored role of conceptual change in education. When one stops to reflect on that for even a moment, it sounds extremely paradoxical. How could conceptual change possibly be ignored by education? Prima facie, it would seem that if students are to be brought from one cognitive state with its associated concepts, beliefs, and ideas to a subsequent cognitive state with its associated concepts, beliefs and ideas, then that change, insofar as it involves changes in concepts at all, ought to be one of *the* central concerns of education. How could one conceive of learning at all without considering conceptual change as an integral part of the whole process?

Strange as it may sound, I will urge, first, that educators, with the possible exception of some cognitive and developmental psychologists, have tried to conceive of learning, teaching, and curriculum planning *without* a proper appreciation of the role of conceptual change and without anything like an adequate account of the phenomenon which I wish to call conceptual change. Second, I will argue that the process of conceptual change is best understood as a process of adaptation, ultimately explicable from an evolutionary standpoint.

© Hugh G. Petrie.
Proceedings of the Philosophy of Education Society, 1976. 117-132.

I

Let me turn to my first major claim—that no satisfactory account exists of the phenomenon I wish to call conceptual change. And let me begin by sketching two examples of the kind of phenomenon I mean. The first comes from the history of science, while the second involves the kind of problem faced almost every day by contemporary science teachers.

Consider the concept of combustion. For a long time people felt that in heating an object one was driving off something so that the products of combustion were something less than that with which one began. Indeed, observation of common cases of burning seemed to confirm this notion. When wood burns, only ash is left. The product driven off came to be called phlogiston and it figured in some very complex theories. Following Priestley and Lavoisier, however, one began to find cases where apparently something is *added* in combustion. When mercury and iron are heated, for example, the resulting product weighs *more* than it did prior to the heating. Gradually our concept of combustion changed to account for these phenomena. It came to be seen as a process of oxidation where oxygen is added to materials. Thus, the concept of combustion changed from one in which essentially something was lost to a concept in which essentially something was gained. The process involved, I call conceptual change.

The second example of the phenomenon I am calling conceptual change involves the teaching of science, in particular explaining Newton's laws of motion. For all of recorded history people have been interested in explaining physical motion and its changes. Why, for example, do various projectiles, rocks, arrows, cannon balls, and so on fall to the earth? Now, according to Newton's laws, a body in motion at a uniform velocity will continue in motion unless acted on by an external force. We all *know* that, if we have taken even a high school physics course. Yet my colleagues in science education tell me that *college* students training to be science teachers consistently seem not to understand that law.

When asked, for example, how far a puck will travel on an infinite, frictionless air hockey table when hit with a certain force as against being hit with twice the force, almost all will say that the harder the puck is hit, the farther it will travel. When asked to explain why, they typically talk about the lighter force "wearing out" first, not being as strong, and so on. Now this idea of "wearing out" is almost a paradigm case of the historical concept of impetus. That is, before Newton, the most advanced physics treated projectile motion as due to the forward motion imparted to the projectile by the means of projection. That power then "resided" in the projectile and was "used up" during the flight of the projectile. Obviously the more power or "impetus" given to the projectile, the farther it would travel.

Clearly, concepts of impetus regarding projectiles not only can, on a gross level, handle the data to be explained, but many contemporary college students, who "know" the right (Newtonian) concepts, still apparently retain large doses of the old impetus notion. I suspect many of us would have to stop and consciously apply the Newtonian concepts to avoid falling into the same trap as the prospective science teachers. The process of giving up the concept of impetus and replacing it with the concept of Newtonian motion in a straight line unless acted upon by external forces is an example of what I mean by conceptual change.

These examples are important because they are of situations where for most people both the original set of concepts and the later set of concepts are more or less accessible. One can appreciate *both* the impetus *and* Newtonian views of motion. Thus such examples offer us a relatively *complete* picture of conceptual change. On the other hand, many others of our current concepts are so firmly embedded in our conceptual schemes by now that we really can't appreciate what it would be like to view the world through the spectacles of the ancestors of our current concepts or through some possible legitimate heirs of those concepts. Thus, I, for one, am not truly capable of seeing how the concept of heat could have ever been associated with the notion of caloric, a colorless, weightless, invisible fluid. Unless I remind myself of the very phenomenon of conceptual change I am here at pains to demonstrate, it becomes incredibly easy to dismiss the phenomenon I am trying to point to.

II

Having given two examples of what I am calling conceptual change I want now to turn briefly to a consideration of just how intimately connected are the problems of conceptual change in the growth of knowledge and the educational problems of curriculum, teaching, and learning. I should make clear that I am speaking of *rational* conceptual change in both the areas of the growth of knowledge and in the area of curriculum. No doubt conceptual changes of all sorts have occurred within the history of humanity and as a result of all sorts of causes.

However, by considering the role of conceptual change in the growth of *knowledge*, I am concerned with conceptual change which contributes to that growth, i.e., with rational conceptual change. Similarly in the field of curriculum, one is concerned not with any old changes of concept which students may undergo, but with those connected to promoting the rational goals of the curriculum. So my investigation has a heavily normative aspect. How can we understand conceptual change as contributing to the rational activities of the growth of knowledge and the furtherance of educational curricula? Thus conceptual change is for me a part of a general theory of rationality—both human rationality in general and educational rationality in particular.

What I want to suggest here is that there is a very strong logical similarity between the general growth of human knowledge on the one hand and the *educational* process conceived as a kind of "growth of knowledge" on the other hand. This similarity is *not*, however, that the growth of human knowledge is a social process, and the educational process is an individual instance of this. Both processes include social *and* individual conceptual change.

I have argued at length elsewhere that with respect to conceptual change the logical position facing the research scientist on the frontiers of knowledge and the beginning science student about to acquire the brand-new (to the student) concepts of the science is virtually identical.[2] Current concepts are inadequate for both. Simple extensions of current concepts will not work. Both require new ways of looking at things. Creativity on the part of each is required, but in both cases, the creativity must be based upon where each currently "is at." Not all attempts at new ways of looking at things will be successful. The scientist may propose many new false hypotheses. The student may misinterpret or misunderstand the new concepts being taught. A kind of social consensus operates on both the scientist's hypothesizing and the student's attempts to understand. In the case of the scientist this comes from his peers. In the case of the student from the general constraints of the curriculum. The process I have called conceptual change is quite central to both the growth of human knowledge in general and to that more particularized growth of human knowledge we call education.

There is, of course, one very important disanalogy between the general growth of human knowledge and education, and that is that in the case of education both the current state of conceptual development of the interested parties *and* the desirable end state of conceptual development are in principle known. Thus the teacher of any given subject can come to know both where the student currently "is at" as well as presumably already knowing the concepts of the subject—where the student should end up. Interestingly, this provides a new and strongly epistemological aspect of the link between concerns for the student's current cognitive state and the desirability of subject matter competence on the part of the teacher.

The situation of the research scientist is somewhat different in this regard. There is no "teacher" (except perhaps "Nature") who knows where the conceptual change ought to be headed. So the scientist must rely on the results of theoretically promising experiments (questions of nature, as so many scientists have put it) for guidance as to the appropriate conceptual changes. The student on the other hand, can question a teacher who is not neutral as Nature is said to be but ideally is actively engaged in helping the student bring about the appropriate conceptual changes. However, I would urge that although this disanalogy should in principle make education a little easier than expanding general human knowledge, it does little to change the similar epistemological predicaments of the student and the scientist when viewed from their perspectives. Thus the phenomenon of conceptual change is of paramount importance in any educational context which involves the acquisition of a relatively new (to the student) set of concepts.

III

What I wish to do in this section is to examine several accounts of conceptual change which seem to be presupposed by much contemporary thinking in education. I will argue that each of these accounts fails fundamentally to capture the phenomenon I have described, and, hence, fails fundamentally to give adequate guidance to educational decisions. I will utilize two criteria of adequacy for a successful account of rational conceptual change proposed by Stephen Toulmin.[3] These are that we must account both for the historical *continuity* of conceptual change and for the *diversity* of concepts we actually find—synchronically and diachronically, at a time and over time. The continuity criterion means that we must account for the principle which rationally links earlier concepts with later ones. The diversity criterion means that we must be able to allow for conceptual diversity without falling prey to the notion that such diversity is simply arbitrary and perhaps radically subjective.

The first account I wish to consider is what I shall call the *atomist* account of conceptual change. On this account there are basic "atoms" of some sort or other out of which all of our concepts can be built. Typically, in an empiricist mode, these atoms are something like sense data which are directly accessible and which can be combined in various ways to build more complex concepts. The *locus classicus* of this view is in the British empiricists; Locke, Berkeley, and Hume are good examples. The concept of a unicorn, so this account would go, is a straightforward combination of the concepts of horn and horse, which concepts in their turn can be analyzed ultimately into simple directly accessible "atomic" concepts (the simple impressions and ideas of the empiricists) which, because they *are* the "atoms" require no further account. The conceptual learning experiments described by Bruner, Goodenow, and Austin in their *Study of Thinking* seem to presuppose an atomist account of concepts.[4] Conceptual change on this view will be nothing more than the recombination of these atoms from one state to another or possibly the addition or subtraction of atoms. The educational analogue of this approach is the emphasis on clearly defining one's learning objectives in advance. Clarity is to be achieved by picking out the "atoms" (whatever they are) which are directly accessible to both student and teacher and then exhibiting how they are to be combined to give the concept to be learned.

Well, what's wrong with this approach? As a start, if the atomist epistemology underlying the idea of clearly defining learning objectives in advance were true, it would follow that *teaching* as normally understood would be unnecessary. If we could only get those atoms identified and clearly exhibited in the particular relations in which they stand for the concept in question, nothing more would be necessary, except perhaps a field trip or demonstration to acquaint the student with the requisite atoms which were not already familiar. The student would transparently *have* the concept. The atoms are already known and the truly successful definition of the concept which is the learning objective shows the student how the atoms are to be combined. What else *could* a teacher do? Those educationists who would dispense with teaching by designing teacher-proof curricula seem to presuppose an atomist account of concepts. The prima facie absurdity of doing away with teaching in education ought to cast serious doubt on the atomist presupposition.

There are other, well-known, objections, however, to an atomist account of conceptual change. Let me just remind you of two. First, the search for the "atoms" seems fruitless. Sense data have *not* been found.

Units of behavior are not unambiguous, and the attempts to make them so seem more to have shown the inherent difficulties of ever finding basic atoms of perception or behavior than to resolve the difficulties. Analysis into smaller and smaller parts serves an explanatory function only in case there is some fundamental part to be found whose explanatory role is clear. With respect to perception and behavior, the context seems to play such a *large* role in even giving significance to the "atoms" that the whole approach seems misguided.

Secondly, a number of philosophers and psychologists and historians have begun to show the incredible extent to which observation is theory-laden, and analogously, the extent to which concepts are belief-dependent. Philosophers have tended to describe the interconnection between belief and concept in terms of the breakdown of the analytic-synthetic distinction. If we cannot sharply distinguish statements true in virtue of their meaning and logical form from statements true in virtue of the way the world is, then we cannot sharply separate concepts considered somehow as connected to the meanings of terms from beliefs about how those concepts actually apply to the world. Psychologists concerned with language acquisition, representation, and use, approach the same point by speaking of our inability to separate our knowledge of language from our knowledge of the world. Ralph Page and I have expressed the significance of this point in another context by noting,

> There may be only one reality, but it does not seem that people must construct only one representation of it, and the choice between the representations people do create must be made without comparing them to unrepresented reality, or reality "itself."[5]

If these two points are true—that there are no explanatorily-favored atoms, and that schemes of representation change as wholes—then conceptual change cannot be accounted for on an atomistic basis. The fundamental reason is that the kind of conceptual diversity allowed by atomism is a combinatorial one while the kind of conceptual diversity we seem to find is a gestaltist one.

The second approach to conceptual change can be seen as a variant of the atomist approach, but it is widespread enough to warrant separate treatment. This is the *operationist* account. On this view every usable concept is ultimately definable in terms of some concrete operation or other. In its most extreme form, concepts *are* nothing more than the operations in terms of which they are defined. Clearly operationism is almost a hall-mark of Anglo-American psychology. Ben Underwood's methodological work represents a fairly pure form of operationism.[6]

Conceptual change on this view is really quite simply accounted for. Concepts change when the underlying operations change. A different operation gives a different concept. Thus one changes from measuring loss of phlogiston to gain of oxidized products. With respect to impetus and Newtonian mechanics, there is a similar change in operations. We change from calculating loss of impetus to explain the falling to earth of a projectile to calculating the attractive force of gravity. Again, the concepts automatically change. Educationally, the emphasis on behavioral outcomes operationally defined, echoes this approach.

I have argued at length elsewhere that such an approach contains enough confusions and half truths to fill a book.[7] However, for now it is probably sufficient to note that this approach, although it clearly admits conceptual diversity—every operation defines a new concept—is totally incapable as it stands of accounting for the *continuity* which is present in conceptual change. On the operationist view conceptual change seems wholly arbitrary and unmotivated. There is no particular reason to change from one concept to another. It just happens. Conceptual change on this view would be discontinuous and unmotivated because to specify a new operation is to specify a new concept. There is no independent common core shared by two operations such that both can be seen as efforts to operationalize the same concept. The theoretical core is to *follow,* not precede the operationally defined concepts.[8] Yet clearly there *are* reasons for conceptual change. It doesn't just happen. The history of the development of any concept amply demonstrates this fact.

Thus in a peculiar way operationism and atomism commit complementary sins. Atomism seems to account for the continuity which persists through conceptual change. It is continuity seen as the continuity of atoms from structure to structure. Yet the conceptual diversity which one actually sees seems unexplained on a view which assumes the same fundamental atoms in all concepts. Thus atomism fails to account for conceptual diversity. Contrariwise, operationism accounts for conceptual diversity with a vengeance. Every operation defines a different concept. Operationism's difficulty is in accounting for the continuity present *in* conceptual change.

A third view of conceptual change I shall call *interpretationism.* It believes there are atoms to be found in concepts, but these atoms are not simply recombined or added to during conceptual change; rather, the atoms are sometimes *reinterpreted*. Thus interpretationism attempts to meet the objection to atomism by providing a more adequate account of the nature of conceptual diversity while retaining atomism's account of continuity.

Thus there is still the concept of burning, but the atoms we thought went into that concept were faulty, we must look again to find the atoms we had misinterpreted originally, perhaps even thinking they were ultimate when they were not, and interpret them correctly. The projectile still falls to the earth but we interpret its fall as due to gravitation rather than the dying out of an imparted force. Educationally, we interpret the operations defining our goals as *indicators* of the goals rather than as identical with them. Most current views of conceptual change are interpretationistic with a dash of operationism thrown in.

Interpretationism is a sophisticated doctrine and, I suppose, elaborated in just the right way, it could be seen as similar to what I am later going to suggest. However, the standard interpretation does *not* seem to me to be correct, and it is that interpretation which I wish to attack. The basic feature of the standard view is that there are atoms or epistemologically (and psychologically) ultimate parts of concepts to be found. A completely developed science of psychology would have finally found these ultimate constituents. Our problems arise in that these ultimate bits are hard to find. If we only could find them, interpretationism would reduce to atomism. Thus, insofar as interpretationism relies on atomism, the arguments concerning the theory-dependence of observation and the inability to find such atoms continue to hold good. The atoms will at best have to be parts of representational systems upon which we agree rather than independently epistemologically ultimate.

But of even more importance, it is just misleading, not to say false, to suppose that we are somehow "given" some sensation or other and then by applying relatively independent mental entities called concepts to this experience, we come up with categorical judgments. The equation

Given + concept (interpretative) = full-blown experience

must be wrong. And the reason for this is the long-standing gestaltist insight that *what* is given depends on the concepts brought to the experience. There are *in principle* no ultimate atoms to be found.

This point is strikingly illustrated by the perceptual ambiguous figures. Consider the duck-rabbit for example.

The ears of the rabbit are not given *as* ears until the rabbit-concept is already applied. The concept helps *determine* the given. The bill of the duck is not given *as* a bill until the duck-concept is applied. (Otherwise that part might be given as ears.) Nor will it do for the interpretationist to say that lines are given and we then interpret them as duck or rabbit. First, the lines lead equally in *either* direction. And further, to see just the lines in situations like this requires a great deal of training. It is, of course, not logically necessary that the epistemologically given be identical with the psychologically given. Thus the interpretationist could simply brush aside this last objection. However, historically the two have been identified, and for good reason. Without *some* kind of independent mark of the epistemologically ultimate, it becomes an empty formalism to claim that such ultimates ground our knowledge.

Nevertheless, the interpretationist account will serve in those cases in which we can agree on *relatively* ultimate atomic constituents for our concepts. The relativity is to the representational schemes we do in fact accept. What we must keep in mind is that sometimes those very representational schemes, or parts of them, will be precisely the subject of conceptual change. What count as atoms depend on the representational scheme and there is no single epistemologically or psychologically favored scheme. However, if one forgets the relativity to a representational scheme, it can look very much as if interpretationism can successfully account for conceptual change.

IV

What, then, would an adequate account of conceptual change look like? The central reason for changing concepts has been taken to be to ascertain most adequately the truth about the world. Our judgments about truth, i.e. our belief systems or theories, are couched in the categorical concepts found in our representational schemes. Traditionally the nature of those representational schemes has been taken as relatively unproblematic. We simply apply them directly to our experience which results in the basic observational data we have about the world. Truth then consists of a correspondence between our beliefs and theories on the one hand and our experience given in the terms provided by our representational scheme.

This neat picture changes drastically, however, when one accepts the thesis of the theory-dependency of observation. Now there is no direct access to the world. Our very representational schemes themselves are intimately bound up with our beliefs and theories, and the influence of the world must be highly indirect, affecting the theories and representational schemes as a whole. Still one can, I think, make sense of what one might call a philosophical concern for the truth.[9] By this I mean simply that we can ask how well our cognitive structures as a whole—representational schemes, theories, basic concepts and beliefs, methodologies for inquiry, and so on—allow us to deal with the world. And we can ask how and under what conditions we ought to make adjustments in our cognitive structures—how we can be rational. But rationality will now be broader than truth-seeking. Truth-seeking is only *one* among many purposes people have for constructing theories, concepts, and representational schemes. How well the rest of human purposes are met must also be considered.

But there is a profoundly important feature in the shift from the relatively straightforward purpose of ascertaining the truth about the world to the much more diffuse purpose of asking how well our cognitive structures as a whole allow us to deal with the world. As Toulmin might put it, the question we must now ask is how well we can say what there is there to say.[10] Both the earlier and later conception of truth, picturing the way the world is versus adapting to it, can be characterized as versions of what I have called the philosophical concern for truth, but the ways they are fleshed out are quite different. In the former case, there is implicitly a single truth there to be discovered and that truth may modify human purposes and reasons but it is essentially independent of them. In the latter case, the representational schemes we use will depend not only on the indirect editing effects of the world *on* our representational schemes, but *also* directly on the purposes we have for constructing these schemes. Human purposes now are seen to be directly relevant to determining truth conditions.

Emphasizing adaptiveness and the role of human purposes in the changing structure of human cognition leads to an extremely important result. The connection between the epistemological concerns of philosophers and the psychological concerns of cognitive development theorists are seen to be extremely closely knit. And this in turn points the direction to an adequate account of conceptual change in terms of adaptiveness. Ideally cognitive development theory tells one about the development of the representational schemes of individuals. What sorts of experiences do they have? How do these experiences change? What does a given mode of experiencing allow a child to learn? What does it keep the child from learning, and so on? Conceptual change on the individual level occurs too, and as it occurs, the child's modes of experience change as well. These changing modes of experience will also entail changing modes of justification. Psychologically *and* epistemologically appropriate justifications for a sixth-grader may well differ from the justifications appropriate for a college sophomore.

Let me illustrate this interdependence of epistemology and psychology in accounting for conceptual change by considering two different kinds of adaptiveness. I shall call these two kinds of adaptiveness, assimilation and accommodation.[11] Assimilation will mean processing our perceptual inputs to make them fit our concepts. Accommodation will mean changing our concepts to fit our perceptual inputs.

The examples with which I began the paper are probably examples of accommodation. We changed the concept of combustion in the light of recalcitrant experience. As one gives up seeing projectiles falling to the ground as the dying out of an imparted force, and rather as the positive action of gravity, one has probably changed one's concept of projectile motion. An example of assimilation would be to see that gravity applies equally to an object in free fall and, for example, a bullet. A bullet fired horizontally over level ground and one dropped from the end of the barrel at the same time would hit the ground at the same time.

Notice, too, that consonant with a developmental theory, someone *could* assimilate for quite some time, even if objectively it seems accommodation is the more adaptive mode. That this actually does occur is borne out in a host of common sense examples from examples of bigotry, to isolated societies such as the Amish, to physically isolated native groups, to students who just don't *see* the point of physics, and, hence, decide it is irrelevant. That this is to be expected also follows on my account of rational conceptual change. For it is a *balance* of selective pressure from the environment—and representations of the environment—on the one hand with human purposes on the other. If selective pressures are kept to a minimum and purposes are strong, even if narrow, then narrow ranges of adaptiveness are certainly to be expected. At the same time the assimilative kind of adaptiveness which does occur within such narrow ranges is also to be expected and will be absolutely astounding in its ingenuity. The ability of narrow-minded groups and individuals to assimilate the wildest sorts of events to their own world-view never ceases to amaze people. I do not mean to imply here that assimilation is necessarily some deficient sort of adaptiveness. On the contrary, most learning *is* probably assimilative. We become better and better at utilizing the basic concepts we have in processing our perceptions. Such examples as the assimilation of Aunt Louise's China dog to a child's concept of dog is probably typical of assimilated learning. It is *very* widespread.

The general phenomenon of conceptual change, conceived as adaptiveness, however, probably covers both assimilation and accommodation, although it is most striking in the accommodation cases. Nevertheless, it is, I think, crucially important to keep assimilation and accommodation separate because I believe there are two radically distinct kinds of mechanisms which account for the two kinds of adaptiveness. Assimilation is, I believe, to be accounted for in terms of control system theory. I have expounded on the role and uniqueness of such forms of explanation in education elsewhere and will not repeat the details here.[12] The crucial point is that a control system or negative feedback model of explanation allows one to handle a *great* deal of variable behavior as being directed towards maintaining the system in some state of dynamic equilibrium *(not rest)*. Thus a fair amount of learning can be understood in terms of refinements and increases in sensitivity of existing homeostatic systems. Surely the difference between the novice and the expert in a sport involves a great deal of such refinement.

On the other hand, when assimilation is no longer possible, accommodation is the remaining developmental alternative. In this case, as Donald Campbell argues forcefully in his, "Evolutionary Epistemology," paper in the Popper *Library of Living Philosophers* volume, the mechanism of adaptation is almost surely evolutionary.[13] I do not mean here that the mechanism directly involves biological evolution, although I do not deny that possibility either.[14] I simply want to emphasize that I do not take `evolution' here to be merely a metaphor. I mean that the accommodative adaptiveness surely will involve the variation, selection, and retention of appropriate trials where these trials need not be of whole organisms.[15]

V

In concluding, let me briefly recapitulate where we have been and attempt to draw a few implications for educational policy, educational research, and educational evaluation.

I have urged that conceptual change is a pervasive phenomenon in human learning, both with respect to the growth of knowledge in general and with respect to the growth of individual knowledge. I illustrated the sort of thing I meant by conceptual change and argued that the situation faced by the beginning student facing a wholly new subject area and the scholar on the frontiers of knowledge is logically very similar. Both need to change their concepts to adapt to their experience. Next I examined three current accounts of conceptual change and found them all wanting. As an alternative I have proposed a view of conceptual change based on adaptiveness *both* to the impact of the world on our cognitive structures *and* on the human purposes we have in utilizing our cognitive structures.

Finally I distinguished two kinds of adaptiveness, which I called assimilation and accommodation. Assimilation occurs when one changes one's experience to fit one's concepts and accommodation when one changes one's concepts to fit one's experience. I asserted that the mechanism underlying assimilation is to be understood in terms derived from control system theory and the mechanism underlying accommodation is blind variation and selective retention, although not necessarily a biological evolution.

With respect to educational policy, one of the central questions has always been how to justify the selection of curricula one finds in the schools. And it seems clear that the burden of justification does lie on those who would intervene in another person's life. Nor does it do anything but push the problem back a step to define "education" as justified intervention. The question then becomes, as our students intuitively seem to sense, "Is what goes on in schools, then, really education?"

Consider then what introducing the impact of human purposes on justifying conceptual schemes means for justifying any curriculum to the individual student. The traditional intellectual and scholarly disciplines surely embody some very general human purposes. But those purposes may *not* be congruent, at least not obviously, with the student's own purposes. Indeed, given the extent to which schemes of representation may vary as a reflection of different purposes, the student may even fail to grasp the significance of the purposes embodied in the intellectual disciplines. At the same time, we seem to have a kind of historical evidence for the general worthwhileness and applicability of the intellectual disciplines. They would probably not have developed as they did if they did not serve a great number of fairly basic human purposes. Thus there is good reason to believe that most, if not all, students will ultimately need something like the knowledge and skills embodied in the standard intellectual disciplines in order to achieve their own developing purposes. How can we bring together the purposes of individual students and the general purposes embodied in the disciplines?

Students, inchoately for the most part, sense the incongruity of their own purposes and the purposes embodied in the disciplines. They then call for relevance and the call is made doubly frustrating to educators by the recognition that although one may be able to justify the intellectual disciplines on grounds of the growth of human knowledge, such a justification may not be appropriate for the individual. That is, one may be able, historically, to show that a *society* needs the knowledge, skills, and techniques embodied in the representational schemes of any given discipline. However, in a highly technologized, post-industrial society *individuals* may *not* need such knowledge to fulfill their purposes. Why should I know physics as the physicists know it when I can buy a transistor radio for less than ten dollars? Why ought I to do sociological inquiry when I can read the latest Gallup poll? Why should I investigate and evaluate political candidates when I can watch Eric Sevareid? And so on.

The point is that by looking *both* at individuals' cognitive schemes and how they have developed *and* at the development of intellectual disciplines we may be able to see the crucial points of conceptual change, compare them, and see how they might justifiably be made congruent. Thus, for example, it may well be that for purposes of general education or satisfying distribution requirements, a first course in college physics needs to be much different from such a course offered to probable physicists. And this does *not* mean merely that the former should be a watered-down version of the latter. Rather it means that potentially a course in the history of physics might much more adequately help achieve the purposes of the non-major—and legitimately so. After all, on this view we have given up the notion that physics ever could, someday, give us *the* picture of reality which might transcend all special purposes for wanting to know about the world. Rather, physics embodies a certain set of purposes legitimate for humanity as far as they go, but by no means eternally fixed and unchanging, and, least of all, necessary for the good life for every individual.

At the same time, given the generality of human purposes embodied in the disciplines, standard curricula are far from arbitrary. Indeed until one tries out the different disciplines, one may not even be able to see what ways experience might change, fruitfully and progressively.[16] Of even more importance, however, the account of conceptual change in terms of adaptiveness gives a framework for considering when and under what conditions it might be justifiable to allow students to pursue curricula different from those to be found in the standard disciplines; namely where their legitimate purposes can be shown to diverge significantly from the norm of human purposes. In this sense, one has the beginnings of an epistemology which might make sense out of much of what is otherwise nonsense written about open education.[17]

Let me turn now to an obviously important area in educational research suggested by my analysis. Conceptual change as a form of learning is radically different from ordinary types of learning, and will require some radically new curricular approaches. How, after all, does one deal with learning where, *in principle,* the student does not initially understand what he is being asked to do? The student, like the research scientist does not know the shape of the accommodation which is appropriate for the task. However, the fact that in the educational process the teacher presumably does know the shape of the individual accommodation needed for the student learning a new discipline gives the student a big advantage over the research scientist. What this means is that curriculum planners and teachers can take advantage of knowing the student's current cognitive state and knowing the accommodation they wish the student to make. They can then intentionally devise experiences to force the accommodation while making the transition as easy as possible. Note, however, that under the hypothesis that concepts are to change fairly radically, and given my rejection of standard modes of interpreting conceptual change, the task facing the curriculum planner is not simply one of putting together the mental equivalent of an erector set. Indeed, I have argued elsewhere that the key pedagogical tool needed is an expanded notion of metaphor.[18]

Felicity Haynes is finishing her dissertation on some of the ways metaphor can bridge the gap between earlier and later cognitive structures, between private and public schemes of justification, and between the logic of discovery and the logic of justification. Basically, however, the notion is obviously promising in that traditionally the main function of metaphor has been to allow us to see a new area more clearly by utilizing familiar lenses in a new way. We start with the student's current cognitive map and ask the student to look at the new area with old tools. The infinite air hockey table begins to help students grasp Newton's laws of motion. Hydraulic models (or metaphors) are useful in teaching electricity and so on. On this view, metaphors turn out to be not only useful heuristic devices in learning, but epistemologically essential to understanding how learning of the conceptual change variety is even possible.

Finally, let me turn to educational evaluation. In particular I will concentrate on evaluating whether or not a student can be said to have undergone a given conceptual change. Consider, first, the process of assimilation. If the educative goal is to get students to assimilate appropriate experience to a given set of concepts, then essentially one is talking about refining or "tuning" existing concepts. For example, a physics student may have minimally grasped the concept of gravitational attraction and the educational task is to get the student to see attraction in such diverse phenomena as free fall, pendula, inclined planes, and so on.

Notice that even in the case of assimilation, once one gets very much more complicated than simple-minded, single-criterion concepts, the idea of actually being able to prespecify all the behaviors which would count as exemplifying the concept becomes slightly ludicrous. Behaviorally-oriented evaluation has really become a classic case of the tail wagging the dog. As one begins to evaluate educational programs, early work with behavioral methods yields some preliminary results consistent with our basic knowledge of what the educational program being evaluated is. But then the behavioral evaluation insists that no other program will even count as educational unless these early gross methods of evaluation are applicable to it.

This reversal exemplifies the deep-running difficulty with behavioral evaluation as ordinarily conceived. Essentially the difficulty is that for all but the simplest concepts and skills, behavioral descriptions are simply too far away from what the agent is really *doing* to give an adequate indication of what is going on. Consider, for example, the evaluation of teaching. We know quite well that good teachers are sometimes supportive and sometimes critical depending on the context. Yet behavioral descriptions identifying critical and supportive behaviors will, typically, count such behaviors as different. The behavioral descriptions, "supportive," and "critical" are simply too far away from what the teacher is *actually* doing, say, "helping the student learn." *That* level of description gives a unifying perspective from which we could see "criticism" and "support" as *both* appropriate in certain contexts.

But the truly revolutionary changes for educational evaluation occur when one moves to the area of accommodation. The basic point here is that from the perspective of the person undergoing the conceptual change, evaluation can ultimately be done only *ex post facto*. If one already knew where one was going with a conceptual change, it wouldn't *be* a conceptual *change*. The point is that the investigative modes, procedures of inquiry, standards of justification, even the very criteria of evaluation themselves are subject to change in any accommodative conceptual change. What counts as good evaluative evidence will be in part dependent on the very conceptual change under consideration. Thus, sensitivity to everything that is going on, portrayal, and other similar types of evaluation are bound to be more appropriate in looking at accommodative change than is experimentally-designed evaluation.

Two quick examples of this dependency. Recall again, the analogy between research scholar and beginning student. For the scholar, the evaluation of his efforts *will* be *ex post facto,* by nature. For the student, the teacher who knows the subject will be able to evaluate successful accommodation in terms of the desired ending cognitive state. On the other hand, evaluation of the student's progress *by the student* will remain *ex post facto*. What this implies is that a certain range of student evaluation of courses designed to promote accommodative conceptual change is *logically impossible* for any student who did not succeed in the course. The best such students could do would be to try to help locate their frustrations so that the teacher could perhaps devise better metaphors for them. And this seems intuitively correct. One just doesn't place much weight on poor evaluation by poor students in courses like physics or philosophy. Until they "get it," at least minimally, they just don't have the tools to evaluate such fields. Where one is dealing with accommodative conceptual change, a peer evaluation of the aptness of the teaching metaphors used is far more appropriate than behavioral objectives.

Second, consider any truly new educational program. From a kind of accountability perspective of accounting for societal funds one can, of course, specify in advance various criteria by which the program will be judged. Such evaluation is familiar. Sometimes it is done well and sometimes poorly. But everyone knows that occasionally such programs really do turn into something which was wholly unanticipated by the designers, and which is nevertheless extremely valuable educationally. It is at least arguable that the result the whole Coleman project has had in casting serious doubts on social science research in educational policy areas is an example of this phenomenon. What actually happened in that process has changed, at least in part, our criteria for educational evaluation. In a very real sense the growing agreement in judgments of the people actually participating in such efforts precedes any statement of criteria which might emerge in terms of which we could evaluate the program. In short, a proper appreciation of conceptual change implies that there is a whole area of evaluation by judgment which logically precedes any evaluation by criteria.[19] This seems to me an area of evaluation not even touched by current work.

There is conceptual change. It is not well accounted for, and when we do come to understand it as adaptiveness, we are going to have to change our concepts of education, evaluation and policy-making fairly radically. Perhaps I have tried to provide some of the metaphors necessary to initiate this change.

Notes

1. I want to thank the members of my seminar on epistemology and cognitive development for the countless times they have stimulated my thinking in this area. In particular, I want to mention Bob Halstead, Bruce Haynes, Felicity Haynes, Graham Oliver, Ralph Page, Martin Schiralli, and Ron Szoke.
2. Hugh G. Petrie, "Metaphorical Models of Mastery: Or, How to Learn to Do the Problems at the End of the Chapter of the Physics Textbook," forthcoming in the *Proceedings of the Philosophy of Science Association*, 1974.
3. Stephen Toulmin, *Human Understanding:* Vol. 1, *The Collective Use and Evolution of Concepts* (Princeton, N.J.: Princeton University, 1972), p. 139.
4. J. S. Bruner, J. J. Goodenow, and G. A. Austin, *A Study of Thinking* (New York: Wiley, 1956).
5. Ralph Page and Hugh G. Petrie, Review of Alan Montefiore (ed.), *Neutrality and Impartiality: The University and Political Commitment* (London: Cambridge University Press, 1975), forthcoming in *Studies in Philosophy and Education*.
6. B. J. Underwood, *Psychological Research* (New York: Appleton Century, Crofts, 1957).

7. Hugh G. Petrie, "Can Education Find Its Lost Objectives Under the Street Lamp of Behaviorism?" in Ralph Smith (ed.), *Regaining Educational Leadership: Essays Critical of PBTE/CBTE* (New York: Wiley, 1975), pp. 64-74.
8. I am indebted for this way of putting the point to Robert Halstead.
9. F. L. Will made this point in a recent seminar on conceptual change.
10. Stephen Toulmin, "The Concept of Stages in Psychological Development," in T. Mischel (ed.), *Cognitive Development and Epistemology* (New York: Academic Press, 1971), p. 38.
11. See, for example, J. Piaget, "The Role of the Concept of Equilibrium in Psychological Explication," in J. Piaget, *Six Psychological Studies* (New York: Vintage Books, 1968). I believe my usage of these terms is the same as Piaget's but I am more concerned with considering them as I have defined them independently of Piagetian theory.
12. Hugh G. Petrie, "Action, Perception, and Education," *Educational Theory*, Vol. 24, no. 1 (Winter, 1974), pp. 33-45.
13. See Donald Campbell's impressive, "Evolutionary Epistemology," in P. A. Schilpp (ed.), *The Philosophy of Karl Popper*, Vol. 14, I and II, *The Library of Living Philosophers* (LaSalle, Ill.: Open Court, 1974).
14. However, some writers on evolutionary epistemology, notably Campbell, do believe that evolutionary adaptiveness is ultimately to be traced to biological evolution.
15. See Stephen Toulmin, *Human Understanding, op. cit.* for a detailed discussion of evolutionary explanation which may yet fall short of relying on biological evolution.
16. Hugh G. Petrie, "That's Just Einstein's Opinion: The Autocracy of Students' Reason in Open Education," in David Nyberg (ed.) *The Philosophy of Open Education* (Routledge and Kegan Paul: London), 1975.
17. One of my students, Robert Halstead, is currently pursuing this avenue. See his, "Evolutionary Epistemology and Open Education," in mimeographed form.
18. Hugh G. Petrie, "Metaphorical Models of Mastery," *op. cit.*
19. Another student, Eric Weir, has cogently presented this case (in manuscript form) in his, "Evaluating Faculty as Teachers Without 'Criteria'," University of Illinois.

[1979, 1993*]
Metaphor and Learning

Metaphor in education

There seem to be two main views of the role of metaphor in education. On the one hand, there is the idea that metaphors are primarily of aesthetic value, with perhaps some secondary utility as heuristic aids. This view concentrates on metaphors along with other linguistic forms, such as analogies, similes, and synecdoche, as figures of speech in literature, especially poetry. The poet's insight is often expressed through metaphor. Occasionally, proponents of the aesthetic value of metaphor also admit that it has some heuristic value in educational contexts outside of literature. For example, some of metaphor's relatives, like analogies and models are often used as teaching aids (see for example, Mayer, 1993; Petrie, 1976). The solar system model of the atom is familiar to high school physics students. But even in such a positive view of the pedagogical value of metaphors, it is usually claimed that although possibly useful and often ornamental, the metaphors and models are not essential to a cognitive understanding of what is being taught and learned. This is at least part of the position held by those whom Black (1993) called the appreciators of metaphor.

On the other hand, metaphors occasionally receive a bad press in education. Metaphors are used when one is too lazy to do the hard, analytic work of determining precisely what one wants to say. Consequently, metaphors encourage sloppy thought. In addition, metaphors can be tremendously misleading. There are a number of different ways in which metaphors can be understood and so the possibility of mistake abounds. If metaphors are eliminated, there will be fewer mistakes. Finally, metaphors and their close cousins, slogans, are often used to

---

Metaphor and learning, by Hugh G. Petrie and Rebecca S. Oshlag (from) Metaphor and Thought, 2nd Edition, edited by Andrew Ortony. Copyright © 1993 Cambridge University Press. Reprinted with permission.

* This chapter is a revision of one by Hugh Petrie that appeared in the first edition (1979) under the same title.

cloud educational issues and reduce complex matters to simple-minded banalities. In short, as has been noted in other connections, metaphors have all the advantage over explicit language as does theft over honest toil (for example, R. M. Miller, 1976). Such views are often held by those whom Black (1993) called the depreciators of metaphor.

Notice that both appreciators and depreciators of metaphors in education tend to agree that the *cognitive* significance of metaphor is severely limited. The main home of metaphor is in poetic insight and any more general cognitive function is ideally better served by explicit analytic language. At best, metaphors may be nice, but they are scarcely necessary for comprehension, communication, or coming to know (but cf. Ortony, 1975).

That view was challenged by Petrie in his chapter in the first edition of this volume where he argued that metaphors had a number of important cognitive roles, in particular, a possibly unique educational role in helping in the acquisition of new knowledge. One thing seems quite clear: in the intervening years, a number of cognitive roles for metaphors have been widely discussed and investigated (see, for example, Williams, 1988, for a summary of work on the cognitive roles of metaphor; Stepich & Newby, 1988, for an analysis of the function of analogies as learning aids within an information processing paradigm; and various chapters in Vosniadou & Ortony, 1989). The cognitive importance of metaphor, especially in instructional settings, has been clearly acknowledged since the first edition of this work. Furthermore, the importance of metaphor for the acquisition of new knowledge is being more and more widely accepted. In this sense, the purpose of the first chapter has already been to some extent fulfilled.

At the same time, despite the explosion of interest in the cognitive functions of metaphor, there remain sharp conflicts over the exact nature and use of metaphor in education. In this revision, therefore, we hope to use some of the work that has appeared since the first edition to augment the core ideas expressed by Petrie some 10 years ago. We are convinced that the major emphases of the earlier chapter are still essentially correct, but we also believe that these emphases can now be more perspicaciously and usefully expressed. Consequently, what follows represents in some cases a significant rewriting of the original chapter. We have tried to keep the basic format and the major conclusions so that the commentary that follows still has point, but we have tried at the same time to incorporate new work and clarify obscure points.

The work on metaphor's cognitive significance since 1979 has proceeded primarily on two fronts. On the one hand, it has been argued that metaphor enables one to transfer learning and understanding from what is well known to what is less well known in a vivid and memorable way, thus enhancing learning. This claim is essentially a psychological one, asserting a connection between vividness, or more precisely, imageability, and learning (e.g., Davidson, 1976; Ortony, 1975; Paivio, 1971; Reynolds & Schwartz, 1983). It is an extremely important result that metaphorical teaching strategies often lead to better and more memorable learning than do explicit strategies.

The memorableness of metaphors can also lead, however, to several undesirable consequences. Not only are metaphors sometimes misleading and misused, we have also learned that on occasion they are taken as literal truth, thereby interfering with the later development of more adequate knowledge. Rand Spiro and his associates (Spiro, Coulson, Feltovich, & Anderson, 1988; Spiro, Feltovich, Coulson, & Anderson, 1989; Spiro, Vispoel, Schmitz, Samarapungavan, & Boerger, 1987), for example, have shown that certain very common and useful analogies and metaphors used in the instruction of physicians come to interfere with later learning and a more adequate understanding of the concepts. Despite these dangers Spiro's suggested solutions do not include eliminating the metaphors, but rather utilizing multiple, cross-cutting metaphors and knowledge sources. Thus, even if we grant the possible misuse and misleadingness of metaphor, especially in advanced learning, and even if we were to assume the goal of making what is learned more explicit, it still appears that metaphors and analogies play a central, even indispensable role in the pedagogical process of acquiring that subject. We call this use of metaphor the pedagogical use.

S. I. Miller (1987), however, in criticizing Petrie's original chapter implicitly distinguished between what we call pedagogical metaphors (or analogies) and theory-constitutive metaphors (see Boyd, 1993; Gentner & Jeziorski, 1993). The former may be useful for the teacher in introducing certain difficult concepts. Theory-constitutive metaphors, however, are integral parts of the very structure of a theory at any given time in its development. All theories contain such metaphors, and their usefulness consists of both their ability to help us learn the theory and their inductive fruitfulness in guiding further research in the theory. They are, for Miller, always to be conceived of as way stations toward a more explicit and literal rendering of the theory.

S.I. Miller's (1987) concern was that educational theorists not ignore the problems of the theory-constitutive metaphors embedded within typical "educational" theories, such as operant conditioning and functionalism. He pointed out, correctly, that it is not always clear how metaphors such as, for instance, "shaping" behavior can be of any practical pedagogical use. It is also important to realize that one can, in principle, look at any theoretical approach and question its metaphors. In short, we can, and sometimes should, examine the theory-constitutive metaphors of educational psychology, physics, or even metaphor comprehension itself.

In a similar vein Reyna (1986) described pedagogic metaphors as a type of functional metaphor used to introduce novel concepts by relating them to familiar concepts. These can be contrasted with technical metaphors which are used to describe abstract concepts in terms of more explicit concepts. This distinction appears very close to the characterization of pedagogic and theory-constitutive metaphor described above. Reyna went further, however, in introducing the distinction between mundane and elite metaphors with the former more easily comprehended than the latter.

For our purposes we wish to lump together as educational metaphors all the various categories of metaphors which are useful for increasing understanding by students. Thus mundane, pedagogic metaphors as well as elite theory-constitutive metaphors can be seen as educational metaphors if they are used by teachers and students to enhance learning. There may even be a category that we would call "residual metaphors" which can function as educational metaphors on certain occasions of their use. These are typically concepts and phrases that may be viewed as literal by people fully familiar with a field, but that would be seen as metaphorical from the point of view of a student just learning a field. The "frames of reference" example in Petrie's original chapter, and repeated here later in abbreviated form, can be seen as an example of a residual metaphor that can have an educational use.

Within the category of educational metaphorical use we wish to focus here on Petrie's original claim that the very possibility of learning something radically new can only be understood by presupposing the operation of something very much like metaphor (see Rumelhart & Norman, 1981, and Vosniadou & Brewer, 1987, for examples of researchers who have been seriously investigating the claim that radically new knowledge requires the operation of metaphor). This is not just the heuristic claim that metaphors are often useful in learning, but the epistemic claim that metaphor, or something very much like it, is what renders possible and intelligible the acquisition of new knowledge.

Plato first posed the problem of the acquisition of radically new knowledge in his famous paradox of the *Meno:*

> You argue that a man cannot enquire either about that which he knows or about that which he does not know; for if he knows, he has no need to enquire; and if not, he cannot; for he does not know the very subject about which he is to enquire. (Plato, *Meno* 80E; Jowett translation)

How *is* it possible to learn something radically new?—a question also raised by Pylyshyn (1993).

There is an educational formulation of the issue raised in the *Meno* paradox. If we assume that we can simply pour knowledge into the heads of students, then we are faced with the problem of how those students can ever recognize what they receive as knowledge, rather than as something to be rote-memorized. If, however, we insist, as current conventional wisdom as well as constructivist psychology (Anderson, 1977; Rumelhart & Ortony, 1977; Schank & Abelson, 1977) would have it, that learning must always start with what the student presently knows, then we are faced with the problem of how the student can come to know anything radically new. It is our thesis that metaphor is one of the central ways of leaping the epistemological chasm between old knowledge and radically new knowledge (see also Petrie, 1976, 1981).

The belief that there is such an epistemological chasm depends on certain presuppositions for which we shall not here argue. Although still somewhat controversial fifteen years ago, these presuppositions are now widely accepted. First, experience is never directly of the world as it is, but is always in part constituted by our modes of representation and understanding, by our schemas, scripts, or mental models. For example, we experience the chairs on which we sit as dense and impenetrable, although they are, physicists tell us, composed of clouds of very tiny particles. Second, most learning consists of processing that which impinges on us in terms of a context of rules or representations. These representations form our modes of understanding. Much learning is thus coming to be able to process our experience in terms of existing contexts and schemas and the relations among them. We learn about the Civil War by seeing it is a war within a nation. Third, however, on some occasions we learn by actually changing our representations. The result of *changed* representations is what we call radically new knowledge. For example, the phenomenon of experiencing something in different ways if approached with a different schema is graphically illustrated by the so-called ambiguous figures. Figure 25.1 can be seen as either a duck or a rabbit. Piaget (1972) noted the distinction between these two kinds of learning by distinguishing between assimilation and accommodation. During assimilation, we learn by changing experience to fit our concepts. During accommodation, we learn by changing our concepts to fit our experience.

Figure 25.1

The problem posed by the *Meno* paradox occurs with accommodation. If understanding and learning involve being able to put that which is learned into a schema, as noted in the first assumption above, then how can we ever rationally come to *change* our schemas? It seems we would either have to presuppose that we already possess, at least implicitly, the schema which renders intelligible the radically new thing we are attempting to learn, or else we would have to admit that the learning of something radically new is arbitrary and subjective. Both alternatives, unfortunately, have considerable precedent in education. What we shall suggest here is a third alternative—that metaphor can provide a *rational* bridge from the known to the radically unknown, from a given context of understanding to a changed context of understanding. The central question for us is "how is radically new knowledge possible?" With the presuppositions noted above, that question becomes "how is rational change of schemas possible?" Finally, these "how?" questions are to be taken in the epistemic and not the psychological sense. In other words, the question is "how is one to make intelligible the acquisition of new knowledge?" not "what are the processes involved?"

Our concern with metaphor is derivative from this central educational concern. We believe an examination of metaphor will show that it does, on occasion, play this crucial epistemic role of rendering the acquisition of radically new knowledge intelligible. We have now learned that there are many devices other than metaphor that serve as a bridge from the known to the unknown (see, for example, Gentner & Jeziorski, 1993; Reigeluth, 1980; Rumelhart & Norman, 1981; Vosniadou & Brewer, 1987). Analogies, models, and exemplary problem solutions also sometimes perform this function and, we believe, in very similar ways to metaphor. The feature that all these have in common is that they invite the use of a familiar rule-governed device for dealing with the material to be learned in ways that require the bending or even breaking of the familiar rules. Metaphor is one crucial way this happens; analogies, models, and exemplars are others. Our purpose is to argue that metaphor, as traditionally understood and as an exemplar of these other types of figurative devices, often plays a central role in the acquisition of radically new knowledge.

Metaphor

There are two issues in the voluminous literature on metaphor that are of particular interest for our purposes. The first is the distinction between comparative and interactive metaphors. On the comparative view of metaphor, what a metaphor does is to say implicitly that two apparently dissimilar things have something in common after all. Thus, in speaking of the "flow" of electricity, despite the obvious dissimilarities between electricity and liquids, it is held that there is a fundamental similarity—they both move in a fluid kind of way. On this view, a metaphor is an implicit comparison, whereas a simile or an analogy is an explicit comparison (Green, 1971); metaphors transfer meaning and understanding by comparison. It should be noted that the notion of a comparative metaphor would not serve to make intelligible the acquisition of radically new knowledge. The problem is that radically new knowledge results from a *change* in modes of representation of knowledge, whereas a comparative metaphor occurs within the existing representations which serve to render the comparison sensible. The comparative level of metaphor might allow for extensions of already existing knowledge, but it would not provide a new form of understanding.

There are problems, however, with attempting to construe all metaphors as implicit comparisons. Consider the example (Haynes, 1975), "Virginity is the enamel of the soul." Is the implicit comparison to be between the positive features of clarity, strength, and protectiveness, or the negative features of rigidity, brittleness, and enclosure? Nothing in the metaphor tells us and only nonlinguistic contextual knowledge of speaker or hearer seems useful. For reasons such as this, many writers have claimed that there is also an interactive level of metaphor. Black says, "It would be more illuminating in some of these cases to say that metaphor creates the similarity than to say that it formulates some similarity antecedently existing" (Black, 1962, p. 37). The interactive level of metaphor is particularly appropriate for our purposes, because if it *creates* similarities, then it could provide the bridge between a student's earlier conceptual and representational schemes and the later scheme of the totally unfamiliar subject to be learned by the student. Interactive metaphor would allow truly new forms of knowledge and understanding to be acquired by the student without presupposing the student already knows, in some sense, that which is being learned.

The discussion so far points to the fact that a metaphor, comparative or interactive, depends on the cognitive scheme presupposed for its understanding. One and the same metaphor can be comparative *and* interactive, depending on the point of view taken. An educational metaphor like "The atom is a miniature solar system" is probably a comparative metaphor from the point of view of the teacher. The teacher already knows both about the solar system and about atoms and is relying on the similarity between them that already exists in our collective understanding. But from the point of view of the student just beginning physics, the metaphor, assuming it is successful, will be interactive. It will (help) create the similarity *for the student.* It provides a way of understanding how the student's existing modes of representation and understanding can be changed through interaction with the new material, even granting that experience is dependent on a particular mode or scheme of understanding.

In the original chapter, Petrie discussed the issue of whether a metaphor can be identified by some set of linguistic features independent of its use on particular occasions. The purpose of the discussion was to address the apparent fact that metaphors are, if interpreted literally, clearly false. The question then became, how do students interpret their teachers when they are uttering falsehoods? In retrospect, this issue is almost certainly a red herring. The common situation in the classroom is that students typically take the teacher to be serious and sincere and when teachers do introduce metaphors, their reports indicate a carefully planned, strategic use of metaphors as part of teaching the new material (e.g., Biermann, 1988a; Marshall, 1984; D. B. Miller, 1988; Whitman, 1975; Zegers, 1983). Given the care and thought that typically go into teachers' use of metaphor, it is unlikely that students will be surprised by pedagogical metaphors. Good teachers know from long experience that certain topics and fields are difficult for students to understand. What happens is that good teachers carefully signal the introduction of something new and the necessity for the students to suspend the normal conversational implications regarding literal truth and falsity. Thus, students will typically try to make sense of the metaphorical utterance, making use of clues that the teacher is serious and attempting to say something important and useful.

Recent work by Glucksberg and Keysar (1990; 1993) throws important new light on the "problem" of the literal falsity of metaphorical assertions. They argue persuasively for the view that, in the final analysis, metaphorical statements are not implicit similes, a view we have at least partially endorsed in arguing for Black's (1962) interactive view of some metaphorical utterances. Instead, they suggest that typical nominative metaphors are class-inclusion assertions in which the topic of the metaphor is assigned to an abstract category referred to by the vehicle of the metaphor. The vehicle thus functions as both the name of the category and as a prototypical example of it. Their example, "My job is a jail," thus receives the interpretation that my job, the topic, is assigned to the class of entities that confine one against one's will, are unpleasant, are difficult to escape from, and so on. The vehicle, jail, is a prototypical exemplar of this new category and serves in this instance as a name for the category.

Glucksberg and Keysar (1990) gave numerous examples of such uses in the language, including a fascinating comment by an Israeli during the war crimes trial of John Demjanjuk who was accused of being "Ivan the Terrible," a sadistic guard at the Treblinka death camp in Poland. Apparently the name Demjanjuk had become a noun in Israel to identify an ordinary person capable of committing unspeakable acts. Thus, it was quite sensible during the trial for an Israeli to say of John Demjanjuk, the defendant, "I know his name is Demjanjuk, but I don't know if he is a Demjanjuk."

This dual role for the vehicle of a metaphor as both prototypical example and name of a newly created class sheds considerable light on how metaphors can actually create the similarity to be noted. As has often been noted, everything is similar to everything else in some respect or other. Some similarities are typically worth drawing attention to and eventually become enshrined as "literal" truth or falsity in terms of typical conceptual schemes. At any point in time, then, the schemas of most people contain certain connections of inclusion, similarity, and relationship, and not others. Good metaphors suggest new connections by picking out an exemplary and well-known example of a certain category, and, by grouping it with a member of another category which is typically *not* related to the metaphor's category, the relevant similarity is created. The particular grouping of existing categories causes the appropriate selection of properties which are to be related by the metaphor in its role as prototypical example of the new class. If such a predication of a prototypical example to the topic continues to make sense, the metaphor may pass into literal truth and become a "dead" metaphor. If it is a bad metaphor, the similarity will not be seen as worth making, or at least not making in that way, and the metaphor will not even be understood or will not catch hold. In either case, the metaphor is anomalous in the sense that a well-known prototypical example of a certain category is connected to parts of our conceptual schemes with which it is not usually associated.

We shall have more to say about how students utilize metaphors to change their cognitive structures. For now we simply want to emphasize that it is the *anomalous* character of an interactive metaphor, anomalous in terms of a student's *current* set of rules for understanding, that distinguishes the way in which metaphors transfer chunks of experience from the way in which literal language or comparative metaphors do. Literal language requires only assimilation to existing frameworks of understanding. Comparative metaphor requires simple extensions of the framework in the light of a more comprehensive framework. Accommodation of anomaly requires changes in the framework of understanding. While these changes in cognitive structures almost certainly fall along a continuum, it is the general requirement of a fairly radical change in cognitive framework that provides the distinction between the ways interactive metaphor and literal language are to be understood. It is this change in framework that secures the importance of metaphor in considering how radically new knowledge is acquired.

Metaphor and the growth of science

The brief description that we have presented of how an interactive metaphor can create an anomaly for a student so as to lead toward changes in cognitive structure bears a striking analogy to Kuhn's description of the workings of science during scientific revolutions (Kuhn, 1970). During the periods of normal science, puzzles and problems are solved by the use of the accepted paradigm of the moment. Occasionally, such problems or disturbances resist current paradigm efforts to solve them, and they become anomalies. The scientist then searches for a new metaphor or model that can remove the anomaly. The main difference between the scientist on the frontiers of knowledge and the student is that in the student's case the metaphor provided by the teacher, if it is a good one, is likely to be more immediately helpful than are the variants tried out by the scientist. Except for a kind of trust in the teacher, however, the student does not really know any more about where he or she will end up than does the scientist. This seems to us to go directly against the educational dogma that one should always lay out in advance for the student exactly what the goals of the learning experience are taken to be. In cases where the goals are to change significantly the student's current cognitive structure, it will not be possible to lay out learning outcomes the student can initially understand. Only metaphorically and ex post facto can the student be brought to understand the goals expressed in the terms and categories of the to-be-learned subject matter.

One of the crucial senses of "paradigm" for Kuhn (1974) is what he called an exemplar. An exemplar is a concrete problem with its solution, which together constitute one of the scientific community's standard examples. Acquiring these exemplars is a critical part of the scientist's training, and they serve the absolutely central function of allowing the student to "apply theory to practice," although, as we shall show, this is a misleading way of making the point. The exemplar is what enables the student to deploy the symbolic generalizations of the theory being learned in particular problem situations. This role is extremely important, because on Kuhn's view, we do not always link up theory and observation statements by means of correspondence rules, nor is there any direct access to the world independent of our theoretical language. In short, having denied a direct perceptual link to the world "as it is," and having accepted the fact that observation is theory-laden, another account of the link between our beliefs and nature must be provided. Kuhn's suggestion is that, in an important sense, exemplars serve this function.

How do exemplars work? Kuhn (1974) gave an extended example of a young boy learning to recognize ducks, swans, and geese by repeated ostensive definition and correction of mistakes. His account went no further than the simple observation that this is indeed how such learning often happens. Kuhn claimed that what the boy learned is not "rules" of application, but rather a primitive perception of similarity and difference. This perception precedes any linguistic formulation of the similarity relations. Can these nonlinguistic similarity relations be spelled out in more detail? If so, perhaps a third alternative, besides direct access and correspondence rules between theory and observation, can be given some plausibility as a way of accounting for the link between observation and nature or between theoretical language and observational language about nature.

What we wish to suggest is that understanding an interactive metaphor includes, as an essential part, activities similar to those involved in acquiring an exemplar. For when a metaphor has effected a change of cognitive structure (where the "rules" of the cognitive structure need not be explicitly formulated or formulatable), the student has a new way of dealing with, describing, and thinking about nature, just as the science student, in acquiring an exemplar, has a new way of deploying symbolic generalizations in nature. Furthermore, if Glucksberg and Keysar (1990) are right in suggesting that metaphors are class-inclusion assertions, with the vehicle of a metaphor being a prototypical example of the new category to be learned, the similarity to the Kuhnian (1974) description of learning new perceptual categories is even more striking. What happens in both cases is that our cognitive structures or schemas are expanded and linked up in different ways through the use of an exemplar of the category being learned. "The atom is a solar system" thus becomes the attribution of the atom to those categories of systems in which there are central bodies around which revolve other bodies with certain forces and relations obtaining, of which the solar system is a prototypical exemplar.

The key to understanding the learning of new categories such as ducks and geese on the one hand, and comprehending metaphors on the other, is that both processes are bound up with *activities* on the part of the student. It is not simply a case of hearing words, understanding them literally, and applying them directly. In both instances it is a case of *acting* in the ecology. For the science student, this is brought out by Kuhn's (1974) insistence that in acquiring exemplars the student requires diagrams, demonstrations, and laboratory exercises and experiments. Even the young boy learning about ducks, swans, and geese is doing something.

He is classifying and being corrected. Of course, language is involved, not as a kind of labeling, but as a prod to *activities* of sorting, classifying, and perceiving similarities and differences. In the case of the metaphors, the activities are again those of classification, building new relationships, testing hypotheses suggested by the new class-inclusion relationship, and the like. We believe our subsequent discussion of interactive metaphor in a pedagogical situation, especially as viewed from the student's perspective, will be directly relevant to Kuhn's claims that exemplars provide the way of understanding how language relates to the world.

Thus the educational functions we are proposing for metaphor are that it does, indeed, make learning more memorable, and that it does, indeed, help move one from the more familiar to the less familiar. But we are also claiming that metaphor is what enables one to pass from the more familiar to the unfamiliar in the sense that it provides a key mechanism for changing our modes of representing the world in thought and language. It provides this mechanism not through a direct labeling, or through explicit rules of application, but rather because in order to understand an interactive metaphor, one must focus one's *activities* on nodes of relative stability in the world. Language bumps into the world at those places where our activity runs up against similar boundaries in diverse situations.

Metaphors and pedagogical content knowledge

Perhaps one of the most influential developments in education since the first edition of this book has been the widespread acceptance of the notion of pedagogical content knowledge (Shulman, 1986; 1987). This is a kind of knowledge that expands on ordinary content knowledge in the direction of those aspects that are particularly germane for teaching the particular content. Shulman has identified two major subcategories of pedagogical content knowledge—first, the most useful forms of metaphors or representations of a subject and, second, the features that render any given topic more or less easy to teach or understand. An example of a powerful metaphor in Newtonian mechanics would be conceiving of the action of objects on each other as if they were a system of billiard balls. With respect to the issue of typical difficulties teachers ought to know about, an example might be the fact that even many college students believe, incorrectly, that if one gave a puck a push on an infinite, frictionless air hockey table, eventually the force imparted to the puck would "wear out" and the puck would stop moving. That is, students tend naturally to hold an impetus theory of motion rather than a Newtonian one. Clearly, both these kinds of knowledge would be very useful for a teacher to know in attempting to teach mechanics.

For our purposes, there are at least two important consequences of the increasing significance of the notion of pedagogical content knowledge. First, it encompasses an explicit acknowledgment of the centrality of metaphor in teaching. We shall return to this feature in a moment. Second, it challenges the traditional conception of learning how to teach. That conception implicitly assumes that one first learns the content of a subject and then one learns general theories of pedagogy. General pedagogical and psychological learning theories are then applied to the content, and, perhaps to the specific students, in order to devise instructional strategies for the content and context in question. Furthermore, it is assumed by many that, on the whole, learning how to teach is more or less content-free. That is, the difference between a physics teacher and an English teacher is believed to lie almost wholly in their respective content knowledge. How to teach it is largely the same for the two and consists of knowing things like how to motivate students, how to structure a lecture, how to manage a class, how to use small group discussions, how to construct grading schemes, and so on.

The concept of pedagogical content knowledge, without denying the usefulness of general pedagogy, invites us to look beyond such principles and focus on the different ways in which content knowledge may be held by both teacher and student. Some of those ways may be more pedagogically useful than others. Some ways of representing that knowledge may make it easier to acquire than others. Some ways of representing a given knowledge domain may be useful for one group of students in one context, but it may be necessary for the expert teacher to have a variety of ways of representing and re-presenting content knowledge so that different kinds of students in different contexts can learn.

Reminding us that there are extremely important pedagogical features specifically connected to the content being taught or learned reinforces the importance of the use of metaphor in education that we have been urging. Indeed, as noted above, Shulman (1986; 1987), has characterized knowledge of the metaphors of a field as one of the key features of pedagogical content knowledge. We would urge that Shulman's metaphors need to include all of what we have called educational metaphors—the theory-constitutive metaphors of the field, the pedagogic metaphors (sometimes they will be the same as the theory-constitutive metaphors) which help introduce students to the field, and even the residual metaphors, those parts of a field which may be viewed as literal truth for people already knowledgeable in the area, but which may involve radically new knowledge for someone just being introduced to the field.

If, therefore, one thinks of the typical student and the typical teacher as each having some sort of conceptual representation or schemas of, say, physics in their minds at any given time, two features stand out. First, as noted earlier, the use of a metaphor, pedagogical, theory-constitutive, or residual, by the teacher may be comparative in that the teacher already knows enough physics to comprehend both the old and new knowledge domains. For the student, on the other hand, the very same metaphor may be interactive, creating the similarity under consideration. Second, the student may well be acquiring one of the constitutive or residual metaphors of the field for the first time.

In the following section we repeat the example from Petrie's original chapter, although in a shortened form. Originally Petrie offered the example as a clear case of the use of a metaphor in a teaching situation. In retrospect, the case does not seem so clear, although as an example of how a given concept that might be viewed as literal truth by those familiar with the field and as a metaphor by students just learning the field, it still seems to us to have merit. We will follow the frames of reference discussion with an example from more recent work which clearly does illustrate how metaphor seems to be the only way in which to overcome student misconceptions.

Educational metaphors: Some examples

One of the interesting features that seems to characterize most people's unreflective concept of motion is that there is no difficulty in deciding whether something is in motion or not. One simply looks and sees. Yet, an essential feature of motion is that it is properly describable only relative to a coordinate system. Where the observer happens to be located when trying to decide whether something is in motion is essential to understanding motion (for simplicity's sake we assume a stationary observer). After noting several examples of motion, one secondary-school science text (Fisk & Blecha, 1966, pp. 217-18) suggested that the reader look at a nearby object, for example, a chair, and decide whether or not it is moving. The authors assumed the answer would be no, and then they pointed out that the chair is on the earth's surface, and the earth is moving, so is not the chair moving after all? The authors were attempting to introduce into the student's conceptual scheme an anomaly analogous to Kuhn's (1974) description of anomalies in the growth of science. Does the chair move or does it not?

Fisk and Blecha had to assume two things about the student; first, that his or her standard unreflective judgment would be that the chair is not moving, and second, that the student knows the earth moves. Without these two assumptions, the attempt to introduce an anomaly into the student's view of the world will fail, for the student will simply reject one of the things he or she is being invited to consider, probably the claim that the earth moves. What this illustrates is that an anomaly will *be* an anomaly only from the standpoint of a conceptual scheme. If the student does not know about the earth's movement, no anomaly will occur. This point illustrates the feature that in order for metaphors to work, at least one of the categories being used metaphorically must be part of the student's conceptual scheme.

Next, Fisk and Blecha (1966) tried to make the anomaly explicit by suggesting that it may seem strange to say a book is both moving and not moving. Here they were relying on the idea that everyone probably finds contradictions anomalous. The theory-constitutive, or, possibly residual metaphor to be used to solve the anomaly is then introduced. The book's moving and not moving seems strange only because the book is being observed from two different *frames of reference*. We take it that the metaphorical term here is "frame of reference."

The authors next define "frame of reference" as "a place or position from which an object's motion may be observed and described" (Fisk & Blecha, 1966, p. 218). It might be objected that "frame of reference" is not a metaphor at all. For the student, however, it may have no literal referent whatsoever. Does it mean that the student is to put up a picture frame and block out part of his or her experience? That would be one "literal" meaning. The point is crucial for pedagogy. A technical term may have a literal meaning for those who understand the subject but be completely metaphorical for the student just learning the subject.

It will be objected that the term was *given* a literal definition and so still fails to be metaphorical. The plausibility of this objection rests on the presupposition that the students have already grasped the notion of *different points of view*—which is, after all, the core of the frame of reference idea. If they have, then "a place or position from which an object's motion may be observed and described" makes sense as referring to different places at the same time with putative observers at those places at the same time. But if the student has not yet grasped the notion of different points of view, then "a place or position from which an object's motion may be observed" may literally mean to the student his or her own place or position. Thus, unless one presupposes that most of the work of grasping the metaphor has already occurred, the "literal" definition may not do the trick at all.

This point can be brought out another way. In order to demonstrate their grasp of the term, the students will have to be able to do things with it. They will have to be able to solve problems, answer questions, in short, to engage in activities guided by the concept of frame of reference. In the current case, those activities are largely confined to thought experiments (as they necessarily must be in most written materials). The student is asked to imagine the chair on the earth's surface, the thought experiment taking on the logical role of activities that help one to triangulate on motion. The metaphor "frame of reference," however that is initially understood by the student, provides the other leg of the triangulation. If thought experiments do not provide sufficient activity for the student to converge on the idea of relative motion, they could be supplemented by actual activities of the same type.

However, the first attempts at convergence may result in fairly gross approximations, and corrections may be needed. Fisk and Blecha (1966) referred back to the chair example and, using frame of reference language, explicitly suggested that the student look at the chair from a position in space near the moon, and as they put it, somewhat hopefully, "You would probably say that the chair is moving because the earth is moving" (p. 218). With the chair—earth example, they are implicitly correcting a possible mistake which they anticipate some students may initially have made.

In addition to the metaphor of frames of reference, the text uses an interesting diagram (Figure 25.2) and a different example to supplement in a perceptual way the new conceptualization suggested by the metaphor. Through the sequence of pictures, Fisk and Blecha (1966) tried to show how important "point of view" is. They took it for granted the student would, if in a spaceship, say the book fell to the floor. By presenting a schematic series of pictures of the spaceship ascending, another anomaly is created, for the floor is also rising. The pictures also illustrate the alternative conceptualization which can solve the problem. The pictures quite plainly *demand* that one take up a point of view outside the spaceship, and it is that "other point of view" that is the point of the lesson. Again the activity is left to thought experiments. *Both* "book falling" and "floor rising" seem appropriate from the point of view from which the pictures are seen. For the students to check out their ideas on such a fairly subtle point provides opportunity for correction and successive triangulations.

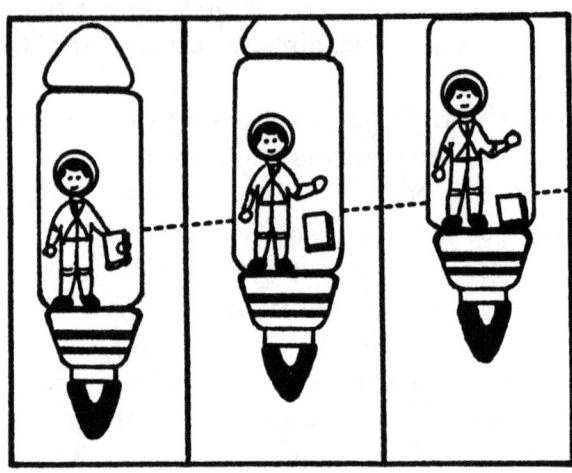

Figure 25.2

The overall point of this example is that, if successful, it has changed the student's conceptual framework in a fairly fundamental way through the use of the term "frame of reference." The notion was given a literal definition in terms of place of observation, but the appropriateness of that definition depended on the *nonlinguistic* ability to take up alternative points of view (another metaphor?), so that place of observation did not simply mean to the student "where I happen to be at the time."

Another, even more dramatic example of an educational metaphor is provided by Joshua and Dupin (1987). They studied the evolution of change-resistant student conceptions and the methods used to overcome these "epistemological obstacles" or "interpretive grids" through which students observe phenomena and then draw conclusions different from those the teachers intended to establish. Of interest were the conceptions of simple direct current held by French students about 12 and 14 years old, in grades comparable to six and eight in the United States.

Four main preinstructional conceptions were identified by Joshua and Dupin (1987) during clinical interviews. The "contact" conception, a simplistic view, emphasized the mechanical contact between the battery and the bulb and was held by relatively few students. Similarly, the "single wire" conception which assumes that electricity travels through just one wire to the light did not enjoy a wide following. The "clashing currents" conception which suggests that two currents leave the battery and supply the bulb but do not return to the battery was held by a majority of the students. Finally, the "current wearing out" view which holds that the current goes one way around the circuit but wears out in its travels through the bulb was maintained by a number of students, particularly the older ones.

Joshua and Dupin (1987) observed the development of these conceptions during a series of activities presented by regular classroom teachers. In the first of two class discussions, students presented their explanations and discussed the various interpretations. No students in either grade put forth the contact conception, perhaps because the teachers' cues and the responses of the other students indicated that more elaborate explanations were sought. The single-wire conception was eliminated based on classroom discussion. The current-wearing-out view was rarely presented, although a "circulatory" conception that did not mention wearing out was voiced. The circulatory view and the clashing current position, held by the majority of the students, remained as competitors at the end of the lesson.

During the second session Joshua and Dupin (1987) observed that students discussed the competing views in small groups but did not typically change positions until a particularly animated class discussion. In the younger group, there was considerable clarification and systematization of views and little change of opinion. Changes did occur in both directions in the older group, but most students simply deepened their initial conception. The circulatory conception was modified and eventually expressed as the current-wearing-out view. This notion had the majority of adherents at the end of the discussion.

Once both the clashing current and current wearing out conceptions were seen as accounting for the phenomenological data equally well, experimentation was used to gain additional information. Joshua and Dupin (1987) reported that the children were not very good at identifying experiments, but they did agree that they needed to know the direction and quantity of current in each wire to decide between the theories. The teachers proposed a method, and the students were required to anticipate the results of the experiment based on their models, carry out the experiment, and draw conclusions. The results of the first experiment failed to support the clashing-current view and were accepted, with some disappointment, by the proponents of that view.

Joshua and Dupin (1987) noted that the teachers had to insist on conducting a second experiment to test the current-wearing-out theory. The supporters of that theory could not accept the possibility of a circulatory model without losses and did not see the purpose of the experiment. When the results demonstrated the same amount of current before and after crossing the light bulb, the younger students were surprised and then unanimously rejected the results, suggesting flaws in either the equipment or its reading. The older students were also surprised by the results but were not so critical of the findings. Most did not reject the experiment, but they also did not accept a result that did not seem logical, that is, the conservation of the electrical fluid in its material form and exhaustion in its energy form. At that point existing student conceptions were simply inadequate to deal with the anomaly presented by the experiment.

To overcome this "epistemological obstacle," Joshua and Dupin (1987) had the teachers employ a "modeling analogy," an analogy operating as a thought experiment. They presented a diagram of a train without a locomotive that operates on a closed-track loop. Workers in a station (the battery) permanently push on train cars going past them, maintaining the movement by tiring their muscles. In their discussion the teachers tried to establish connections between the analogy and the "without loss" conception. The older students gradually began to grasp the analogy whereas the younger children grasped only the connection between the battery and the workers with the remaining connections, such as current-and-train and wires-and-tracks, introduced by the teachers. Joshua and Dupin found that the doubts and criticisms of students in both classes decreased immediately once it was accepted that the battery wears out but the current is the same along the circuit.

Students then became interested in testing the limits of the analogy. They wanted to know what would happen if the tracks were cut. Joshua and Dupin (1987) reported that the students hypothesized, based on the analogy and contrary to their normal experience, that the bulb would stay on temporarily like a train-car derailment. Clearly they were trying out the new metaphor (model) and correcting its implications with thought experiments and actual activities.

Joshua and Dupin (1987) concluded that the students' incorrect conceptions (from the point of view of physics) were, nevertheless, clearly used as a rational basis for their reasoning and evaluative behavior. Implicitly illustrating Shulman's (1986; 1987) notions of what should be included in teachers' pedagogical content knowledge, Joshua and Dupin argued that for instruction to be effective it must take into account and counteract the natural conceptions that students have, conceptions that have a great capacity for logical, if incorrect, adaptation to experience. In short, students do try to account for new experience using their existing schemas, and only if faced with an anomaly and a new way of conceiving of the anomaly, something very much like a metaphor, can they acquire radically new knowledge.

What we believe these examples illustrate is the kind of convergence of thought and activity that can lead through a succession of iterations from a given way of conceiving a situation to a radically different way of conceiving it. In most cases of learning, bringing to bear thought and action in their ordinary literal guises is all that is necessary to resolve the problematic situation. The kind of learning that goes on in such cases is what we have called ordinary learning. It is primarily the utilization of existing cognitive structures to deal with our experience. In other cases involving the use of a metaphor that may be comparative from the point of view of the student, the process is only slightly more complicated. The larger cognitive structure in terms of which the implicit comparison makes sense is already possessed by the student and is straightforwardly brought to bear and the similarities noted. This may be what was going on, for example, as the French students rejected the single-wire conception during the class discussions. In cases like the frame of reference example and the final use of the train analogy in the electricity case, however, neither ordinary learning nor a simple extension of cognitive frameworks allows the student to deal with the problematic situation. In such cases, a change of cognitive structure, or as we have called it, the acquisition of radically new knowledge, is necessary.

The continuum briefly sketched above, from ordinary learning to understanding comparative metaphors to the structural changes consequent on construing an interactive metaphor (probably with a number of other steps between), is very important. The continuum illustrates that one need not be consciously aware of anything so esoteric as metaphors or the need for radical change in one's cognitive structure. In this sense, the process of construing a metaphor is, as Searle (1993) says, a natural extension of ordinary thought and activity. Educationally, we can start with what we know, and by an iteration of triangulations of thought and activity on our experience, end up with radically new knowledge.

Let us now try to analyze this process of acquiring radically new knowledge by means of triangulation on the world. We call the first component of the process the "anomaly step." This consists subjectively of the student's perception of the situation as problematic enough to require a minimal amount of cognitive activity. Objectively, the activity will need to be of a sufficient magnitude to require a change in cognitive structure, although as we have noted, the student need not be aware of this. During the students' initial encounter with the electricity example, they were totally unaware that the result would require a radical restructuring of their schemata.

Assuming, then, that attempts at assimilating the problematic situation to ordinary cognitive structures have failed, or would not work, and there really is an anomaly in the technical sense of a problem requiring a change of cognitive structure, the second step, which we shall call "providing a metaphor" begins. In the typical educational situation the most important source of an alternative classification for the student is the metaphor provided by the textbook or teacher. Yet because the schema suggested by the metaphor has not yet been applied by the student to the material to be learned, the learning process will necessarily be interactive for the student. The metaphor is a guide in that it essentially says, "Look at and deal with this new situation as if it were like one you already know about." Thus, the second step in the process of understanding a metaphor is conceiving of possible variant classifications as if they were like what is already known, so as to create new class-inclusion relations in Glucksberg and Keysar's (1990; 1993) sense.

The third step in the process is actually acting in the world and observing the results. An interactive metaphor is not going to succeed unless activity takes place. The activity is guided by the metaphor. On the first trial, one can conceive of the activity taking place as if the metaphor were literally true. Recall the "train derailment" hypothesis in Joshua and Dupin (1987). One behaves in the new area as one would have behaved in the area in which the metaphor is literally true. The point of the metaphorically suggested activity is to see if it will remove the anomaly. Does the teacher respond positively to the students' papers, questions, and examination answers? Do experiments turn out as predicted? Do thought experiments make sense? If not, do the responses indicate that the activity was close?

Notice that each of the steps thus far mentioned—recognizing an anomaly, conceiving the problematic situation as if the metaphor were literally true of it, and acting on the situation in those terms—provides a triangulation on the problematic situation. The anomaly step focuses on the situation in terms of the existing framework and characterizes the situation in existing framework terms. The metaphor suggests a new characterization of the situation, and activity in the situation provides a third perspective. Of course, some of the steps may be combined, as when a good metaphor creates the anomaly because it is literally false and simultaneously creates a new characterization that can guide subsequent activity.

The fourth step in construing an educational metaphor is the correction of the activity. Typically, the first activity carried out in terms of the metaphor's suggested conceptualization will not be quite right; yet, from the point of view of the anomaly, it will show promise. Much of the problem in the electricity case in Joshua and Dupin (1987) was removed once the students could see how something could wear out, the workers' muscles, while something else could remain the same, the train cars traveling around the track. At the same time more had to be done, especially with the younger students.

The teacher, typically, will provide such a correction, and a new activity—close to, but slightly different from, the original—will be attempted. The corrected activity provides a triangulation of the problematic situation and can be compared to the original metaphorical conception, beginning to show in what sense the ultimately correct conceptualization will differ from and in what sense it will be the same as the initial metaphorically suggested conceptualization. The corrected activity also shows the extent to which the anomaly is being removed by the corrected activity and its evolving conceptualization. It is essentially the iterative process of triangulation of conceptualization and activity, powered by the perception of remaining anomaly, that enables the students gradually to change conceptual schemes to accommodate totally new experiences. And it is the reaching of a final equilibrium of conceptualization and activity that is the test of the success of the metaphor, not whether we can explain its meaning (R. M. Miller, 1976).

The same steps are present when one considers metaphor from the external point of view of the teacher, but they look somewhat different. The teacher must pay special attention to both the initial intelligibility to the student of the metaphor and the appropriateness of activity for triangulation on the new material. The teacher must also consider the ecology in which corrections leading ultimately to a reflective equilibrium in the student take place. For in triangulating on the new material, the student will stop when conception and activity have combined to remove the anomaly. Recall that among the students studied by Joshua and Dupin (1987) those who held the current-wearing-out conception were quite unwilling to have their theory tested. They thought they had removed all the problems through disproving the alternative clashing-current conception. There are equilibria that do not match the collective understanding the teacher is trying to impart, and those are the ones that must be avoided. They tend to be the typical mistakes made by students in any given subject and were referred to above in describing Shulman's (1986; 1987) discussion of pedagogical content knowledge. Probably the best way, in general, to avoid these mistakes is to provide an ecology rich in opportunities to apply the student's newly established equilibrium of conceptualization and activity. For if the student's triangulation is just a bit off, it is more likely to become apparent to the teacher dealing with a variety of cases. This is what the teachers in Joshua and Dupin's study essentially did and it is also the strategy advocated by Spiro, Coulson, Feltovich, and Anderson (1988) in dealing with the misconceptions and oversimplifications of the medical students they studied.

The power of metaphor as one of the ways of intentionally bringing about conceptual change should now be apparent. The teacher presumably has a grasp of both the student's current ways of structuring his or her experience and the conceptual structure as it is found in the material to be taught. The teacher can, therefore, choose instructional metaphors that will serve to remove incipient anomalies for the student, as well as suggest initial conceptual guides to removing the anomalies. Furthermore, activities for the student can be chosen with a view to guiding the successive triangulations of thought and action toward the material to be learned. Students can learn something radically new without metaphors, but only if their variant conceptualizations serve the same function as metaphors—providing new ways to look at old material.

Returning for a moment to Kuhn (1974), the importance of his exemplars or concrete problem solutions is now apparent. For these are the *activities* that provide one of the crucial legs in the triangulation of conceptualization and activity on the subject area. Such problem solutions are indeed how the science student learns to deploy the disciplinary matrix in dealing with the world. The four-step process we have outlined of anomaly, metaphor, activity, and correction can be seen as explicating Kuhn's ostensive definition as an activity in which the student must *construct* the experience to which the definition is to apply. At the same time, if the process of learning a new paradigm is at all like what we have described as the process of a student coming to change conceptual frameworks through the operation of interactive metaphor, then the process of paradigm shift is both intelligible and intelligent. It is intelligible as an iteration of triangulations of thought and action on the world. It is intelligent in that it proceeds *from* the rules of reasonableness currently held by the scientist/student at any point in the historical process.

How are metaphors used in education?

Teachers' anecdotal data support Petrie's original and our continued contention that teachers can use metaphors to bring about structural changes in the cognitive apparatus of students. They also suggest that metaphors can be employed to promote changes in the affective characteristics of students who are learning unfamiliar subject matter which they perceive as irrelevant. Moreover, they indicate that *student* production of metaphors can result in important changes in students' cognitive structures and affective responses. What follows are, first, illustrations of how the components of anomaly, provision of metaphor, acting in the world, and correction of activity are carried out in the real world to effect cognitive change in students. Although the steps frequently overlap in the classroom, and divisions between one step and another are, to some extent, arbitrary. Next is an account of how teachers use metaphors to make unfamiliar material more interesting and relevant to students. Last are indications of how teachers assist students to produce metaphors and descriptions of the kinds of changes in knowledge and meaning that are brought about through metaphor production.

Cognitive change

The use of metaphors and analogies has been reported in the teaching of major subject areas such as biology, business, chemistry, geometry, literature, physics, political theory, psychology, and statistics. Teachers use metaphors to teach concepts that students ordinarily find difficult to learn through factual presentations because the concepts are unfamiliar or complex (Biermann, 1988b; D. B. Miller, 1988; Zegers, 1983). The concepts may be derived from concrete experiences that the students have not had or they may describe abstract realms where direct sensory experience is not possible. For some students, abstract, nonvisual knowledge may be a radically new *type* of knowledge to learn, a type of knowledge they may be unprepared to learn (DiGiovanna, 1987; Garde, 1987; Licata, 1988). Although an occasional analogy may be described as appropriate for middle school students (Allen & Burlbaw, 1987), metaphors are most often used in high schools and colleges. Students at these levels are frequently expected to rather rapidly learn unfamiliar, abstract, and complex concepts, and teachers are more likely to find that conventional methods—examples, arguments, and the drawing of inferences—are insufficient.

Teachers become highly proficient at producing and providing metaphors. DiGiovanna (1987), for example, presented suggestions for other teachers in selecting and using analogies. Like DiGiovanna, teachers are careful to draw vehicles from domains of knowledge that the students have already acquired, especially from important and familiar aspects of students' lives such as school, sports, social relationships, food, and money (Best, 1984; Licata, 1988; Marshall, 1984; Poskozim, Wazorick, Tiempetpaisal, & Poskozim, 1986). Best stated that she had identified domains of knowledge from which to draw analogies for various types of concepts related to political theory. Furthermore, she reported routinely viewing aspects of everyday activity as vehicles for her subject matter. "The technique can be learned. The more that I use it, the easier it gets. Now I see metaphors for my subject matter everywhere I look" (p. 168).

Teachers provide metaphors within contexts that support comprehension. They ensure that students are familiar with the vehicle or direct attention to aspects of the vehicle that students are to use to structure new information (Allen & Burlbaw, 1987; Whitman, 1975). Teachers assist students to visualize metaphors and analogies by describing the vehicle in some detail and directing students to imagine the topic as that situation (Biermann, 1988a; Garde, 1987; Last, 1983). Additionally, teachers present analogies through diagrams, demonstrations with concrete objects, or student participation in actively representing the metaphor (Ball, 1987; Bonneau, 1987; Kangas, 1988; Kolb & Kolb, 1987/88). The provision of a metaphor can be an extended and dramatic affair. D. B. Miller (1988) taught the concept of the interrelationship between genes and experience in the development of organic structure and behavior using the model of a cooking demonstration. The demonstration involved flour (the genetic base), different cooking methods (experimental factors), and different food items (the developmental outcomes).

Teachers typically direct student activity and further guide student thinking. They encourage their students to discuss the metaphor (Cavese, 1976; DiGiovanna, 1987; Garde, 1986), develop examples and solve problems based on the metaphor (Laque, 1978), apply the metaphor in new situations (Polyson & Blick, 1985), or attempt to extend the metaphor past the point where it begins to break down (Marshall, 1984). Teachers may require students to examine limitations of analogies to encourage exploration of the new concept (Licata, 1988), ensure that students will not later be misled by the analogy (Biermann, 1988a), or provide the teacher with feedback on whether the analogy and the related concept have been understood (Webb, 1985).

Correction of activity can occur throughout instruction. Teachers determine whether a given metaphor is being understood based on students' questions, arguments, applications, and various affective responses (Best, 1984; DiGiovanna, 1987; Polyson & Blick, 1985). The extended explanations and demonstrations provide ongoing opportunities for students to compare their construals of the metaphor to that of the teacher and correct accordingly. Students who attend to the remarks of classmates may also compare their own characterizations to those of their peers, and, depending on the responses of others, either maintain their own construals or bring theirs in line with those of the others.

Teachers' use of conferences and professional journals to communicate their successful employment of metaphors points to the development of an area of pedagogical content knowledge. Since 1980, for example, *The Journal of Chemical Education* has featured a collection of applications and analogies designed to assist students understand difficult concepts. The reports also suggest that teachers often perceive the use of metaphors as advisable, if not necessary, when presenting certain new concepts. With the exception of Joshua and Dupin (1987), noted above, empirical research on classroom use of metaphor has generally been directed toward demonstrating the effectiveness of metaphor-based or analogy-based instruction relative to conventional instruction (e.g., Burns & Okey, 1985; Evans, 1988), determining the effectiveness of various types of analogical learning aids (e.g., Bean, Searles, Singer, & Cowen, 1990), or testing competing theories of how and why metaphors are effective (e.g., Evans & Evans, 1989; Simons, 1984). (See Zeitoun, 1984, for a model for teaching scientific analogies that attempts to incorporate research on learning and instruction.) These studies indicate that metaphors aid in the acquisition of knowledge and that lessons employing metaphors or analogies are more effective than conventional methods. Empirical research is needed to identify situations in which metaphors are clearly necessary and how metaphors make possible the acquisition of new knowledge in classroom settings.

Metaphors as motivators

Teachers have indicated that the learning of unfamiliar and abstract concepts is further complicated by affective characteristics of students. Students who are unaccustomed to learning abstract material may dislike dealing with it (Best, 1984), or students may simply not be interested in learning material they perceive as far from their own lives or as difficult (Polyson & Blick, 1985). Some students may have had experience in learning complex and abstract concepts in their own fields of study, but they may find such concepts in other domains uninteresting or irrelevant (Biermann, 1988b; DiGiovanna, 1987). Yet such teachers and others (Hirsch, 1973; Marshall, 1984) have noted that students become interested in learning difficult concepts that are presented through metaphor or analogy.

Haynes (1978) argued that the educational power of metaphors comes from their capacity to bridge the gap between the teachers' rational knowledge and the lack of knowledge of the student by drawing from the shared experiences of the students and teacher. The metaphors or common examples serve the cognitive function of shared rules, but, because they draw from the experiential base of the students, they also include aspects of knowledge that are vivid, emotive, and experiential. This often tacit knowledge assists students to understand new knowledge in their own terms and gives metaphor a dimension of meaningfulness.

The meaning that the vehicle has for the students may not necessarily have been considered during production, but it can have a positive impact on learning. To assist her students read a manometer, a U-shaped instrument for measuring difference between gas pressure and atmospheric pressure, Garde (1986) compared the mercury levels in the two sides of the tube with children on a seesaw. The lower mercury level was associated with the heavier child on the seesaw, the child who made the seesaw go down. Garde reported the effectiveness of the analogy in assisting students to learn to read the manometer, but she also observed that the reference to the seesaw elicited memories of more carefree days and the discussion of the weight of children held the high school students' attention because of their concern with body weight.

Metaphors also permit teachers to provide or transmit meaning for unfamiliar or abstract concepts. Marshall (1984), for example, reported the use of an anthropomorphism that involved telling students they would meet many new "friends" during the year to help them learn. Students initially found it artificial and somewhat embarrassing to refer to unit factors in chemistry as friends, but later in the year, when Marshall would become engrossed in presenting a new concept, it was the students who asked if what was being taught was a new friend, something helpful and nice to have around. Haynes (1978) stated that she transmits her values associated with philosophy by telling her students the study of philosophy is like an orange—requiring effort to remove the tough, bitter covering but sweet and nourishing once one is inside. Hirsch (1973) used behavioral and social phenomena strategically as vehicles for analogies to present physics concepts to liberal arts students who regarded science as boring.

Metaphors also enable teachers and students to share meaning. A metaphor used by Polyson and Blick (1985) to present concepts in experimental psychology to students who typically found such concepts boring and difficult operated by construing basketball as a psychology experiment. A basketball game was presented as a means of testing a hypothesis concerning which team is better in the various mental and psychomotor skills required by the game. The teachers used this analogy during the season when the intercollegiate basketball team at their university had a very successful year and was of great interest to both faculty and students. Although this metaphor clearly established the similarity of structure between the experimental method and rational aspects of basketball games, it would seem to have drawn from emotive aspects of knowledge such as those related to the uncertain outcomes and comparisons of performance that most likely were encompassed in the meaning of basketball games for students and teachers that year and, presumably, to an extent, in the meaning of the experimental method for the two psychologists.

Polyson and Blick's (1985) construal of a basketball game as a psychology experiment is an anomaly in that unlike the typical nominative metaphor the topic, basketball game, is the known situation and the vehicle, psychology experiment, is the unknown situation. Although the students knew nothing more about basketball games immediately after the provision of the metaphor than they did before it was presented, the statement probably did signal the students that they were going to learn about something in which they were interested.

This unusual metaphor can also be analyzed in terms of Glucksberg and Keysar's (1990; 1993) presentation of a metaphor as a class-inclusion statement. *Psychology experiment,* as the vehicle, is the prototype of a category that encompasses certain scientific testing of human performances, a concept that was probably as unfamiliar to the students as was psychology experiment. But whatever psychology experiment was to the students, it was no longer irrelevant because it had become a means of learning about something that they wanted to know about (Polyson & Blick, 1985). The metaphor established the similarity between basketball games and psychology experiments and from that point, the teacher's explanations could be directed toward clarifying just what the similarities were.

The metaphor enabled the development of a new schema for basketball games. During the year, students acquired substantially new knowledge about basketball games, and by the end of the year, they could view them in a very different way. Their concept included much of what they had already known about the games, but it now included the rational structure of psychology experiments as well. Basketball games perceived as psychology experiments would include knowledge related to rules, procedures, the players, and so on. Certain vivid and emotive properties of basketball games such as the high rate of physical energy, mass expression of emotion, and the gymnasium would have been selected out, but aspects of basketball games as highly interesting events involving competition and important but uncertain outcomes would have remained as properties of the topic.

The grouping of the topic and vehicle created similarity between them. This is because the similarity of grouped objects is both causal and derivative (Glucksberg & Keysar, 1990). The ground of the new category was provided by both the topic and vehicle specifying the category in which they had joint membership. As the prototype of the category to which the topic and vehicle belonged, psychology experiments exemplified basketball games and had those properties of basketball games that comprised the ground. Both category members included knowledge related to outcomes and comparing performance. In addition, basketball games had an experimental base from which to draw. This base most likely would have included a high level of such aspects of affect as interest, importance, and excitement. These aspects would have been correlated in the real world with outcomes and performances and would have thereby constituted part of the ground of the category established by the metaphor.

Although for a number of teachers (e.g., Ball, 1987; Garde, 1986) metaphors are useful simply because they enable the presentation of rational structures of concepts, teaching metaphors do have a dimension of meaningfulness or what Carroll and Thomas (1982) called "emotional tone" for teachers to draw on. Under Black's (1993) view, teachers' metaphors are resonant with implications that are highly familiar and meaningful to students. The reports of teachers such as Best (1984), DiGiovanna (1987), and Marshall (1984) suggest that it may be necessary for teachers to make use of this aspect of metaphor when students perceive new concepts as irrelevant.

Student production of metaphors

Although metaphor *comprehension* may make it possible for students to acquire radically new knowledge, metaphor *production* requires some knowledge of the topic. Even though analogies can be produced based on nothing more than the recognition of some similarity in salient properties between a known system and a relatively unfamiliar system (Vosniadou, 1989), the production of metaphors and analogies is more typically used as an instructional device when students know enough about a situation to at least tentatively identify salient elements, conceptualize their relationships, and then search for a similar, familiar situation (e.g., Licata, 1988). Thus, production is used in the mastery rather than the initial acquisition of new concepts. Production of analogies has been employed to assist students to analyze literary works (McGonigal, 1988) and to apply social studies concepts (Wragg & Allen, 1983). Metaphor production can also be used in the acquisition of procedural knowledge. Skills in technical and expository writing have been taught through metaphor production (Catron, 1982; Wess, 1982) as have reading comprehension skills (Kuse & Kuse, 1986).

At this point, our concern is with the production and explication of metaphors for the purpose of communicating conceptual knowledge. Catron (1982) taught advanced science students to produce metaphors in writing for lay readers. His students could already write knowledgeably about scientific subject matter. Their task was to view reality from outside the concepts and theory-constitutive metaphors of their own fields and select concepts from everyday life onto which to map their knowledge.

Sunstein and P. M. Anderson (1989; see also P. M. Anderson & Sunstein, 1987) taught college freshmen to use metaphors to write about unfamiliar science topics. In representing a connection between their personal experiences and a scientific model these students learned writing skills and acquired scientific knowledge. In spite of the differences in the two sets of students and the complexity of their production tasks, the components of anomaly, provision of metaphor, activity, and correction were evident in the instruction described by their teachers.

Each of the components was presented in all four assignments described by Catron (1982). His students did not initially perceive themselves as lacking skills in transmitting their subject matter, and Catron generated disturbances to enable students to perceive the discrepancy between their use of jargon and the strategies needed to convey information to nonscientists. For example, in the first assignment, he required students to describe an unfamiliar object to a general reader, leading them to view artifacts as lay readers. Catron provided examples of metaphors and other figures of speech, often from scientific and technical writings, to demonstrate the technique of their creation or to illustrate their effectiveness. The discussion and writing exercises that ensued permitted students to act on the information and to have activity corrected.

Sunstein and P. M. Anderson's (1989) students perceived an anomaly of substantial magnitude in that they did not regard themselves as competent in writing or knowledgeable about science topics. During prewriting activities students read and analyzed essays by noted science writers and completed worksheets designed to assist them to investigate a topic and compare it to something else. These activities provided the students with examples of the use of metaphor and enabled them to produce a simile or analogy.

Students' early drafts consisted of extended analogies or similes. After students successively revised their work and identified an audience, Sunstein and P. M. Anderson (1989) encouraged them to think metaphorically about their subject matter. They were led to brainstorm, redraft, and shift from communicating new information to representing experience metaphorically. Writers drew more and more directly from their own experiences as they reconceptualized the topic as the vehicle. One student employed a "verb pass" which involved using verbs from the vehicle to describe the topic. Another presented facts following the order produced by the metaphor (P. M. Anderson & Sunstein, 1987). Metacognitive evaluations by students indicated greater scientific understanding and awareness of the production process.

Similarly, Wess (1982) taught students to use analogies to generate ideas for writing, to discover resemblances, and to see the world in a different way. Although Wess analyzed student writing and retrospective data in terms of a process of inquiry framework of preparation, incubation, illumination, and verification, the four components are evident. The anomaly was represented by Wess' presentation of the assignment. Provision of metaphor was accomplished by the presentation of one of his own analogical essays and a description of the creation process. Students' retrospective essays indicated an awareness of a disturbance followed by considerable activity directed toward determining a topic, an audience, and their roles as writers. Students reported that they regarded the assignment as very difficult and that they initially reviewed class examples, reflected, discussed possible topics with others, and engaged in brainstorming. Topics were identified during these activities or simply came to mind later during diverse and unrelated activities. Once a topic was identified and writing began, correction took place throughout the writing process. Although students wrote about subjects that were relatively familiar, many reported that they gained a great deal of new knowledge about the topic, and some reported affective changes as well. Student comments also indicated a sense of rightness about the analogies that they eventually explicated. Teacher verification provided external correction.

Production studies demonstrate that students can be taught to produce and explicate metaphors. They also indicate the importance of providing examples of metaphors and their use and of teaching the process of production, either through relatively structured activities or a portrayal of the process. The assignment creates the anomaly, and the examples and the information on the process suggest to the students a schema that incorporates information on what will correct the situation as well as the cognitive and behavioral acts that will result in a solution. This schema guides student production and is corrected through that activity.

These studies also demonstrate that production enables the acquisition of knowledge, even for students who are already knowledgeable about the topic. It also makes possible the integration of new knowledge with previous knowledge and thereby aids in thinking about the topic as a member of the same class as the vehicle. Production of metaphor supports both cognitive and affective changes. Sunstein and P. M. Anderson's (1989) data, in particular, indicate the importance of metaphor as a form of language in inducing these changes. The language form required their students to imagine the topic as if it were the vehicle and apparently led the students to draw directly from their experience. In short, students can learn to produce metaphors and thereby form new connections, view things in a different way, and generate explanations.

Conclusion

We have tried to sketch the educational centrality of metaphor for bridging the gap between old and new knowledge. Educational metaphors need to be viewed from two perspectives—that of the student and that of the teacher. From the latter standpoint, the metaphor may look like a concealed analogy, but what the teacher must never forget is that from the student's point of view, in those cases in which the metaphor really is to effect a cognitive change, it will *not* be merely an analogy.

Metaphors in education have traditionally been viewed as occasionally heuristically useful but essentially ornamental, and sometimes as downright pernicious. We have argued that metaphors are essential for learning in a number of ways. They may provide the most memorable ways of learning as well as critical affective aids to learning, and thus be our most efficient and effective tools. But further, they are epistemically necessary in that they seem to provide a basic way of passing from the well known to the unknown. Such a formulation is somewhat misleading, however. The crucial use of metaphor is in moving from one conceptual scheme with its associated way of knowing to *another* conceptual scheme with *its* associated way of knowing. Finally, and of suggestive importance for current philosophy of science, it seems that the activity phase of understanding metaphors has much in common with the use of exemplars—concrete problem solutions—in providing an alternative to immediate observation as one of the crucial legs for triangulating our theories and observations on the world.

One new feature in the recent literature is of particular interest to a discussion of the educational implications of metaphor. Metaphors appear to be construable as class-inclusion statements where the vehicle serves as both a prototypical exemplar of the category being predicated and as a name for that category. This feature shows how metaphors can be used to create an anomaly for the student, how they can provide a new view of the situation, and how they can be judged better or worse without having to speak of "metaphorical truth"; rather, one judges them as more or less successful in suggesting a fruitful new way of organizing our schemas.

Our positive account of how educational metaphors work contains four steps. First, an anomaly is "created" for the student, often through the fact that good teachers know where students tend to have problems with the material to be learned and also know the best pedagogical metaphors and the core theory-constitutive metaphors in the subject in question. Consequently, teachers often introduce new material with metaphors *assuming* that standard factual presentations would create an anomaly for the students. The metaphor provides one leg of a triangulation by suggesting a way of looking at new, unknown material as if it were old, known material. In addition to the new view, opportunity to be active with the new material is critical. This activity may be either directly experiential or may take the form of thought experiments. In either case the activity, the acquisition of nonlinguistic similarity relationships, is essential in providing the other leg for triangulating on the material to be learned. Corrections of initial triangulations and iterations of the whole process provide a mechanism whereby eventually the student's understanding of the material to be learned and his or her manner of acting on the material provide a triangulation that is significantly different from where the student began, and significantly like the triangulation enshrined in the disciplinary, collective understanding of the material, justifying our claim that the student has learned something radically new. The metaphor has been successful, not when we can say what it means, but when the triangulation allows the student to make judgments similar to those of experts in similar specific cases.

Empirical work on the use of educational metaphors in the past decade tends to confirm the centrality of metaphors in acquiring new knowledge as well as being consonant with the four-step analysis we have provided. This work has also suggested an increased importance for the affective characteristics of metaphor and has begun to examine the role of student-produced metaphors in knowledge acquisition by the students. Thus, understanding the process involved in construing metaphor is what makes intelligible the ability to learn something new while admitting we must always start with what and how we already know.

References

Allen, M., & Burlbaw, L. (1987). Making meaning with a metaphor. *Social Education, 51,* 142-143.

Anderson, P. M., & Sunstein, B. S. (1987). *Teaching the use of metaphor in science writing.* Paper presented at the 38th annual meeting of the conference on College Composition and Communication, Atlanta, GA. (ERIC Document Reproduction Service No. ED 281 204.)

Anderson, R. C. (1977). The notion of schemata and the educational enterprise: General discussion of the conference. In R. C. Anderson & R. J. Spiro (eds.), *Schooling and the acquisition of knowledge.* Hillsdale, NJ: Erlbaum.

Ball, D. W. (1987). Another auto analogy: Rate-determining steps. *Journal of Chemical Education, 64,* 486-487.

Bean, T. W., Searles, D., Singer, H., & Cowen, S. (1990). Learning concepts from biology text through pictorial analogies and an analogical study guide. *Journal of Educational Research, 83,* 233-237.

Best, J. A. (1984). Teaching political theory: Meaning through metaphor. *Improving College and University Teaching, 32,* 165-168.

Biermann, C. A. (1988a). Hot potatoes - high energy electrons: An analogy. *The American Biology Teacher, 50,* 451-452.

Biermann, C. A. (1988b). The protein a cell built (and the house that Jack built). *The American Biology Teacher, 50,* 162-163.

Black, M. (1962). Metaphor. In M. Black, *Models and metaphors.* Ithaca, NY: Cornell University Press.

Black, Max (1993). *More about metaphor.* In Ortony, A. (ed.), *Metaphor and thought,* Second Edition, NY: Cambridge University Press

Bonneau, M. C. (1987). Enthalpy and "Hot Wheels" - An analogy. *Journal of Chemical Education, 64,* 486-487.

Boyd, Richard (1993). *Metaphor and theory change: What is "metaphor" a metaphor for?* In Ortony, A. (ed.), *Metaphor and thought,* Second Edition, NY: Cambridge University Press

Burns, J. C., & Okey, J. R. (1985). *Effects of teacher use of analogies on achievement of high school biology students with varying levels of cognitive ability and prior knowledge.* Paper presented at the 58th annual meeting of the National Association for Research in Science Teaching, French Lick Springs, IN. (ERIC Document Reproduction Service No. ED 254 431.)

Carroll, J. M., & Thomas, J. C. (1982). Metaphor and cognitive representation of computing systems. *IEEE Transactions on Systems, Man and Cybernetics, 12,* 107-116.

Catron, D. M. (1982). *The creation of metaphor: A case for figurative language in technical writing classes.* Paper presented at the 33rd annual meeting of the Conference on College Composition and Communication. San Francisco, CA. (ERIC Document Reproduction Service No. ED 217 470.)

Cavese, J. A. (1976). An analogue for the cell. *The American Biology Teacher, 38,* 108-109.

Davidson, R. E. (1976). The role of metaphor and analogy in learning. In J. R. Levin & V. L. Allen (eds.), *Cognitive learning in children.* New York: Academic Press.

DiGiovanna, A. (1987). Making it meaningful and memorable. *The American Biology Teacher, 49,* 417-420.

Evans, G. E. (1988). Metaphors as learning aids in university lectures. *Journal of Experimental Education, 56,* 91-99.

Evans, R. D., and Evans, G. E. (1989). Cognitive mechanisms in learning from metaphors. *Journal of Experimental Education, 58,* 5-19.

Fisk, F. G., & Blecha, M. K. (1966). *The physical sciences.* River Forest, IL: Doubleday.

Garde, I. B. (1986). An easy approach for reading manometers to determine gas pressure: The analogy of the child's seesaw. *Journal of Chemical Education, 63,* 796-797.

Garde, I. B. (1987). An analogy for soluble and insoluble mixtures: Sand and magnetic iron filings. *Journal of Chemical Education, 64,* 154-155.

Gentner, Dedre and Jeziorski, Michael (1993). *The shift from metaphor to analogy in Western science.* In Ortony, A. (ed.), *Metaphor and thought,* Second Edition, NY: Cambridge University Press

Glucksberg, S., & Keysar, B. (1990). Understanding metaphorical comparisons: Beyond similarity. *Psychological Review, 97,* 3-18.

Glucksberg, Sam and Keysar, Boaz (1993). *How metaphors work.* In Ortony, A. (ed.), *Metaphor and thought,* Second Edition, NY: Cambridge University Press

Green, T. (1971). *The activities of teaching.* New York: McGraw-Hill.

Haynes, F. (1975). Metaphor as interactive. *Educational Theory, 25,* 272-277.

Haynes, F. (1978). Metaphoric understanding. *Journal of Aesthetic Education, 12,* 99-115.

Hirsch, G. O. (1973). *Social examples in teaching physical concepts.* Paper presented at the 21st national convention of the National Science Teachers Association, Detroit, MI. (ERIC Document Reproduction Service No. ED 093 598.)

Johsua, S., & Dupin, J. J. (1987). Taking into account student conceptions an instructional strategy: An example in physics. *Cognition and Instruction,* 117-135.

Kangas, P. (1988). A chess analogy: Teaching the role of animals in ecosystems. *The American Biology Teacher, 50,* 160-162.

Kolb, K. E., & Kolb, D. K. (1987/88). Classroom analogy for addition polymerization. *Journal of College Science Teaching, 17,* 230-231.

Kuhn, T. S. (1970). Reflections on my critics. In I. Lakatos & A. Musgrave (eds.), *Criticism and the growth of knowledge.* Cambridge University Press.

Kuhn, T. S. (1974). Second thoughts on paradigms. In F. Suppe (ed.), *The structure of scientific theories.* Urbana: University of Illinois Press.

Kuse, L. S., & Kuse, H. R. (1986). Using analogies to study social studies texts. *Social Education, 50,* 24-25.

Laque, C. F. (1978). *Mathematical designs for teaching and learning composition* Paper presented at the 29th annual meeting of the conference on College Composition and Communication, Denver, CO. (ERIC Document Reproduction Service No. ED 159 719.)

Last, A. M. (1983). A bloody nose, the hairdresser's salon, flies in an elevator, and dancing couples: The use of analogies in teaching introductory chemistry. *Journal of Chemical Education, 60,* 748-750.

Licata, K. P. (1988). Chemistry is like a . . . *The Science Teacher, 55,* 41-43.

Marshall, J. K. (1984). Classroom potpourri. *Journal of Chemical Education, 61,* 425-427.

Mayer, Richard E. (1993). *The instructive metaphor: Metaphoric aids to students' understanding of science.* In Ortony, A. (ed.), *Metaphor and thought,* Second Edition, NY: Cambridge University Press

McGonigal, E. (1988). Correlative thinking: Writing analogies about literature. *English Journal, 77,* 66-67.

Miller, D. B. (1988). The nature-nurture issue: Lessons from the Pillsbury Dough boy. *Teaching of Psychology, 15,* 147-149.

Miller, R. M. (1976). The dubious case for metaphors in educational writing *Educational Theory, 26,* 174-181.

Miller, S. I. (1987). Some comments on the utility of metaphors for educational theory and practice. *Educational Theory, 37,* 219-227.

Ortony, A. (1975). Why metaphors are necessary and not just nice. *Educational Theory, 25,* 45-53.

Paivio, A. (1971). *Imagery and verbal processes.* New York: Holt, Rinehart, and Winston.

Petrie, H. G. (1976). Metaphorical models of mastery: Or how to learn to do the problems at the end of the chapter in the physics textbook. In R. S. Cohen, C. A. Hooker, A. C. Michalos, & J. W. vanEvra (eds.), *Proceedings of the Philosophy of Science Association, 1974.* Dordrecht, Holland: D. Reidel.

Petrie, H. G. (1981). *The dilemma of enquiry and learning.* Chicago: The University of Chicago Press.

Piaget, J. (1972). *The principles of genetic epistemology.* New York: Basic Books.

Plato. (1937). *Meno.* In B. Jowett (ed. and trans.), *The dialogues of Plato.* New York: Random House.

Polyson, J. A., & Blick, K. A. (1985). Basketball game as psychology experiment. *Teaching of Psychology, 12,* 52-53.

Poskozim, P. S., Wazorick, J. W., Tiempetpaisal, P., & Poskozim, J. A. (1986). Analogies for Avogadro's number. *Journal of Chemical Education, 63,* 125-126.

Pylyshyn, Zenon W. (1993). *Metaphorical imprecision and the "top-down" research strategy.* In Ortony, A. (ed.), *Metaphor and thought,* Second Edition, NY: Cambridge University Pres

Reigeluth, C. M. (1980). *Meaningfulness and instruction: Relating what is being learned to what a student knows.* Syracuse, NY: Syracuse University, School of Education. (ERIC Document Reproduction Service No. ED 195 263.)

Reyna, V. F. (1986). Metaphor and associated phenomena: Specifying the boundaries of psychological inquiry. *Metaphor and Symbolic Activity, 1,* 271-290.

Reynolds, R. E., & Schwartz, R. M. (1983). Relation of metaphoric processing to comprehension and memory. *Journal of Educational Psychology, 75,* 450-459.

Rumelhart, D. E., & Norman, D. A. (1981). Analogical processes in learning. In J. R. Anderson (ed.), *Cognitive skills and their acquisition.* Hillsdale, NJ: Erlbaum.

Rumelhart, D. E., & Ortony, A. 1977. The representation of knowledge in memory. In R. C. Anderson, R. J. Spiro, & W. E. Montague (eds.), *Schooling and the acquisition of knowledge.* Hillsdale, NJ: Erlbaum.

Schank, R. C., & Abelson, R.P. (1977). *Scripts, plans, goals and understanding.* Hillsdale, NJ: Erlbaum.

Shulman, L. S. (1986). Those who understand: Knowledge growth in teaching. *Educational Researcher, 15,* 4-14.

Shulman, L. S. (1987). Knowledge and teaching: Foundations of new reform. *Harvard Educational Review, 57,* 1-22.

Simons, P. R. J. (1984). Instructing with analogies. *Journal of Educational Psychology, 76,* 513-527.

Spiro, R. J., Coulson, R. L., Feltovich, P. J., & Anderson, D. K. (1988). Cognitive Flexibility Theory: Advanced knowledge acquisition in ill-structured domains. In *Proceedings of the tenth annual conference of the Cognitive Science Society.* Hillsdale, NJ: Erlbaum.

Spiro, R. J., Feltovich, P. J., Coulson, R. L., & Anderson, D. K. (1989). Multiple analogies for complex concepts: Antidotes for analogy-induced misconceptions in advanced knowledge acquisition. In S. Vosniadou & A. Ortony (eds.), *Similarity and analogical reasoning*. Cambridge University Press.

Spiro, R. J., Vispoel, W. P., Schmitz, J. G., Samarapungavan, A., & Boerger, A.E. (1987). Knowledge acquisition for application: Cognitive flexibility and transfer in complex content domains. In B. K. Britton & S. M. Glynn (eds.), *Executive control processes in reading*. Hillsdale, NJ: Erlbaum.

Stepich, D. A., & Newby, T. J. (1988). Analogical instruction within the information processing paradigm: Effective means to facilitate learning. *Instructional Science, 17,* 129-144.

Sunstein, B. S., & Anderson, P. M. (1989). Metaphor, science, and the spectator role: An approach for non-scientists. *Teaching English in the Two-Year College, 16,* 9-16.

Vosniadou, S., & Brewer, W. F. (1987). Theories of knowledge restructuring in development. *Review of Education Research, 57,* 51-67.

Vosniadou, S. (1989). Analogical reasoning as a mechanism in knowledge acquisition: A developmental perspective. In S. Vosniadou & A. Ortony (eds.), *Similarity and analogical reasoning*. Cambridge University Press.

Vosniadou S. & Ortony A. (1989) (eds.), *Similarity and analogical reasoning*. Cambridge University Press.

Webb, M. J. (1985). Analogies and their limitations. *School Science and Mathematics, 85,* 645-650.

Wess, R. C. (1982). A *teacher essay as model for student invention.* Paper presented at the 33rd annual meeting of the conference on College Composition and Communication, San Francisco, CA. (ERIC Document Reproduction Service No. ED 217 478.)

Whitman, N. C. (1975). Chess in the geometry classroom. *Mathematics Teacher, 68,* 71-72.

Williams, P. S. (1988). Going west to get east: Using metaphors as instructional tools. *Journal of Children in Contemporary Society, 20,* 79-98.

Wragg, P. H., & Allen, R. J. (1983). Developing creativity in social studies III: Generating analogies. *Georgia Social Science Journal, 14,* 27-32.

Zegers, D. A. (1983). An urban example for teaching interspecific competition. *The American Biology Teacher, 45,* 276-277.

Zeitoun, H. H. (1984). Teaching scientific analogies: A proposed model. *Research in Science and Technological Education, 2,* 107-125.

[1979]
Against "Objective" Tests: A Note on the Epistemology Underlying Current Testing Dogma

One of the common-sense distinctions in educational testing is between "objective" and "subjective" tests. The former category includes true-false, multiple-choice, mathematical problem solving, matching, and the like, while the latter comprises essay exams, rating scales, interviews, and ordinary observation of students. Clearly all of these tests are used at one time or another and for one purpose or another, and most educators grant that all of them have a place (e.g., Gronlund, 1976, p. 144 ff). The important thing is to choose the most appropriate test for measuring the intended student learning. Objective tests tend to be better at measuring knowledge of facts and the possession of certain kinds of definite problem-solving skills, while subjective tests tend to be better for assessing a student's ability to organize, integrate, and express ideas in an effective way. At least, that seems to be the conventional and even textbook wisdom.

And yet, it is tempting to view subjective tests as somehow inferior to objective tests. Subjective tests seem to enjoy a sort of second-class citizenship. They're all right for certain purposes if you can't do any better, but one gets the distinct impression that objective tests are preferred whenever possible. It almost seems that it would be preferable to have a flute student's performance scored by a machine rather than by Rampal; we just haven't been able to build the right machine yet.

Where does this rough-and-ready, yet terribly influential, distinction come from? I suspect the source is primarily from the method of scoring. Anyone, even a machine, can score an objective test, while scoring a subjective test requires human judgment, and somehow, we have come to believe that mindless, mechanical procedures are to be preferred to judgment. How has such a paradoxical situation come about?

First published in: Mark N. Ozer, (ed.), *A cybernetic approach to the assessment of children*, 1979, 117-150.

One answer is that mechanical evaluation and scoring techniques can give a more dependable basis for making educational judgments than can individual teachers. The individual is believed to be biased and undependable while the machine is unbiased and consistent. And consistency, replicability, and lack of bias must be included in any concept of objectivity. That is why we prefer mechanically scored "objective" tests to judgmentally scored "subjective" tests. A test is "objective" when format, if not content, creates a decidable procedure for determining right and wrong answers.

Subjective and objective

Two senses of objective and subjective

There are, however, a number of confusions in this way of looking at the distinction between subjective and objective tests. Michael Scriven (1972) has noted one of these. He suggests that the pair "subjective-objective" marks two very different distinctions. On the one hand, the pair refers to the distinction between biased and unbiased. In this sense "objective" is the important term because it indicates the central concept. One is objective when one attempts to keep bias, from whatever source, from creeping into observation, analysis, and argument. It is a term closely associated with logical, rational, intelligent procedures. Being "objective" in this sense means guarding against the intrusion of whims, prejudices, social or class bias, and simple wishful thinking into our knowledge-seeking activities.

When contrasted with "objective" in the sense of unbiased, "subjective" takes on the sense of bias. Of course, one important source of bias is our personal opinions, wants, and desires. We often wish the world were other than it is, and we allow such personal bias to interfere with a rational, objective assessment. We are being "subjective." Notice carefully, however, that although a person's own feelings and internal subjective states may be an important *source* of bias, when used in opposition to "unbiased," "subjective" *means* "biased." It takes its meaning from this opposition and not from the fact that much bias happens to come from personal internal states.

On the other hand, the pair "subjective-objective" can also refer to the distinction between the personal and the interpersonal (or extra-personal). In this case, "subjective" is the important concept because a trait, or state, or feeling is subjective only if it is a trait, state, or feeling which belongs to a single person. My hunger is a subjective feeling in this sense as is my fondness for chocolate-covered peanuts.

When paired with "subjective" in the sense of personal, "objective" takes on the meaning of interpersonal. We could, for example, determine how many people like chocolate-covered peanuts as well as determining the specific gravity of water. Notice with this pair that although it may be relatively easy to be objective (unbiased) about interpersonal affairs, when I say that a given feature is objective (interpersonal) I do not automatically rule out the possibility that there may be bias involved in certain determinations of the feature. The average difference between black and white I.Q. scores may be objective (interpersonal), and yet the procedures for determining this may be highly biased. Highly sophisticated statistical techniques have been developed for dealing with the interpersonal, yet unless the interpersonal is also unbiased, the statistical treatment merely quantifies the bias.

Problems arise when these two *different* distinctions get confused. A common confusion is to *automatically* assume that "bias" and "personal" are always present together. Once this mistake is made, it would seem logically impossible to be objective (unbiased) about things that are internal to a given person. And yet a moment's reflection shows this to be patent nonsense. Although we are not always unbiased about our own feelings, sometimes we are, and in many cases the most objective (unbiased) report about some internal state would be our own subjective (personal) introspection of that state. While the feeling of "love" may lead to *other* unreliable conclusions, it would be absurd to suggest that we really need an inter- (or im-) personal machine to tell us when we are *in* love. In short, in some cases the best way to be objective (unbiased) about certain things is to be subjective (personal).

Now I suspect that it is the confusion I have just noted that gives some substance to the feeling that subjective tests are second-class citizens. For subjective tests tend to emphasize personal judgment, and if we are implicitly confusing the personal with the biased, then clearly objective tests are preferred. Recognizing that there are at least two distinctions at work should help immeasurably in removing the temptation to label tests which rely on subjective (personal) judgments as inevitably inferior.

Another confusion also operates, however, and in perhaps an even more interesting way. This is the case in which the "unbiased" is automatically assumed to be present together with the "interpersonal."

It is true that *one* way, although clearly not the only way, of avoiding bias is in some instances to rely upon interpersonal rather than personal judgments. The justification of political democracy depends on this fact. Empirical science also relies on the interpersonal to a large extent, but only properly trained scientists are allowed to "vote." Science is, fortunately, not left to majority rule.

Interpersonal agreement

I believe that this conflation of "unbiased" and "interpersonal" leads to much of the conventional "wisdom" regarding subjective and objective tests in education. I am referring to the pressures to move away from bias by moving toward tests that stress the interpersonal. Machine scoring turns out to be the most inter- (im-?) personal method we have. If we can agree on mechanical scoring procedures, we have effectively eliminated any hint of personal judgment. Furthermore, this "wisdom" is not all foolishness because in many instances a move to the interpersonal *is* a way to reduce bias. But it will not automatically do so, as is suggested when one mistakenly confuses objective (unbiased) with objective (interpersonal).

I want to look more closely at the role of the interpersonal in avoiding bias. Speaking of evaluation techniques that are clearly interpersonal, Gronlund (1976, p. 4) says, "It is not intended that the use of evaluation techniques replace the thoughtful judgments of teachers, but rather that they provide a more dependable basis for making such judgments." One picture which springs to mind upon reading such a passage (and such remarks are commonplace) is that somehow the goal, were it only attainable, would be to find techniques to replace teachers. If only the measurement techniques were available, whenever there was a conflict between teacher and technique, the technique would win.

If that is the picture we are supposed to assume, it is wrong on historical, methodological, and conceptual grounds. It is wrong historically because the development of any evaluative technique, from packaged standardized test to informal classroom quiz, reflects the initial teacher(s) judgment about the appropriateness and validity of the technique. The development of I.Q. tests is a perfect case in point. Items were constructed, selected, and, at least initially revised with reference to the ability of these tests to match paradigm teacher judgments about "bright" and "slow" children. Only after we were quite confident that the evaluation techniques embodied teacher judgments, did we occasionally let the techniques override teacher judgment in borderline cases.

Methodologically, the idea that techniques take precedence over personal judgment is also a mistake. When one tries to develop formal, comprehensive techniques for dealing with, for example, valid arguments in logic, or gravitational attraction in physics, the judgments about paradigm cases of valid arguments and gravitational attraction are prior to the general techniques. These individual cases are what our techniques have to deal with, and if the techniques don't deal with them, we modify the techniques, not our judgments of which are paradigm valid arguments. Only much later in the development of formal, systematic procedures do we occasionally allow the techniques to legislate over an individual judgment in a borderline case. The teacher can accept a few cases where the standardized test ranks students differently from teachers' professional judgments, but imagine the uproar if the test systematically turned teacher judgments upside-down! Would the tests obviously be a more dependable basis for making educational decisions in such a situation than teachers' professional judgments? Not at all!

The picture of technique always overriding judgment is wrong conceptually as well. I have argued elsewhere (Petrie, 1971) that Wittgenstein shows convincingly that agreement in judgments is the logical basis for giving sense to formal, systematic sets of rules or techniques in any area of human activity. Basically, the idea is that agreement in judgments is to be interpreted as similar ways of acting in similar situations and in the end, we train people into these modes of behavior. We do not have any direct access to reality as it is, we can only deal with the representations we construct of it. What prevents wholesale subjectivism is that we must act in the world and our representations must be such as to allow us reasonably effective action or else the representations will be weeded out in an evolutionary way. The important point for my purposes is that, logically, action in the world precedes static, formal representations of the world.

Specifying educational outcomes

However, even if one grants the logical priority of active judgment over static, formal technique, the requirement of agreement in judgment upon which evaluative techniques must be based might be seen as pushing one toward more "objective" (interpersonal) modes of testing. Indeed this is one way to read much of the current literature on testing. The emphasis, for example, on the use of educational outcomes as the basis for designing evaluation instruments is a case in point. We are urged to specify the desired outcomes of education, then design our teaching and testing procedures accordingly. We are not supposed simply to think of the "ground to cover" in a course, for that tells us nothing about what difference we believe the covered ground will make to our students. "How should they be changed as a result of our teaching?" is the question to be asked, not, "What processes are they going to undergo?" For the latter question is pointless without at least an implicit answer to the former.

One should specify intended educational outcomes; however, not just any specification will do. Some general instructional outcomes are too general. What are we to make of the goals of becoming a good citizen or knowing biology? The point seems to be that we don't have agreement in judgments about such vague, general goals, and so to be objective in both the interpersonal and lack of bias sense, we need to specify these goals in more detail so we can reach agreement in judgment.

The injunction to define educational outcomes in behavioral terms can then be seen as a way to meet the demand for agreement in judgment. If we cannot always agree when we have educated a good citizen, perhaps we can agree when a student has voted, or has paid taxes and surely those are part of being a good citizen. Or are they? One could conceive of situations where paying taxes, e.g., a poll tax used to discriminate, would *not* be an act of good citizenship.

Similarly, if one could not agree on what being a good biologist is, perhaps one could agree on instances of "classifying such and such as bacteria," and those are part of being a good biologist. Or are they? Would that rule out the recent discovery of a third form of life previously classified as bacteria?

The point here is that there seem to be two forces operating. One force pushes toward more and more specific, more and more atomistic specifications of educational goals in an effort to reach agreement in judgment. However, the other force seems constantly to be reminding us that the "atoms" we thus identify are the kind they are by virtue of fitting into an overall contextual gestalt that cannot be analyzed away into atoms. Only the context gives meaning to the atoms themselves.

This principle is strikingly illustrated by the perceptual ambiguous figure. Consider, for example, the martini-bikini (top of next page). Is the circle, for example, a boy's navel or an olive dropping into a martini glass? We don't decide whether we have a martini by first deciding whether the circle is an olive. On the contrary, the circle becomes an olive in the context of the martini.

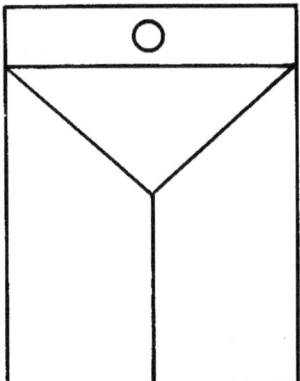

One of the thrusts of behavioral objectives in education thus could be interpreted as an attempt to find context-free atoms upon which we could all agree. If that were possible, then we could specify educational goals in terms of these context-free atoms, plan teaching strategies to lead to these goals, and devise evaluative techniques which would tell us unambiguously when these goals had been reached. How successful has this program been?

I cannot enter here into a detailed polemic against behavioral objectives in education. However, I think it can be fairly stated that the program for finding context-free atoms of behavior related to course content to specify our educational goals and upon which all can agree has not achieved an outstanding success. (See, for example, Smith [1975] for typical critiques.) It is nevertheless instructive to trace the standard moves of the behavioral objectives program.

I have already noted that the first stratagem is to insist that certain common-sense educational goals, such as "training a good biologist," are too vague and general. More specific goals are urged and their specification in behavioral terms is likewise demanded. And to the kinds of objections of context relativity I have raised, the response is to break down the goals into more and more discrete parts. So instead of "identifying bacteria," we might get "identifying such and such a slide as bacteria of a certain kind."

My suspicion is that one can always invent a context, no matter how specific the goals become, that would render the goal inappropriate (Petrie, 1977). I believe it is a general feature of our understanding that specific goals only make sense in presupposed contexts, just as the specific parts of ambiguous figures only make sense given one or the other of the gestalts. If this is so, then the behavioral program of basing agreement on judgments in atomistic elements of teaching and learning is misguided. We will have to presuppose the agreement in judgment on the larger gestalt to even give sense to what the bits and pieces are. However, to pursue that here would take me too far afield. Instead, I shall try to untangle some of the educational presuppositions and implications which seem to accompany the behavioral objectives bandwagon.

Problems with educational outcomes

In the first place, the effort to gain agreement in judgment seems to lead not only to smaller and smaller units of analysis, but also to more and more standardized cases. People are more apt to agree on the tried and true than on the novel and creative. And despite some lip service paid to using learning in new situations (transfer of training), the innovative response is clearly discriminated against when one goes the behavioral objectives route.

This tendency to exclude innovative, yet appropriate, responses from the category of "right answer" is heightened by the injunction to tie one's testing procedures closely to specific and behaviorally stated goals. Gronlund (1976, p. 60) says, "The final step in the evaluation process is to *select or develop evaluation instruments that provide the most direct evidence concerning the attainment of each specific learning outcome.*"

Later in that chapter he indicates that, for example, if the learning outcome says the student is to "supply" a definition, then the test must ask the student to *supply,* rather than, e.g., *select* a definition.

The advantage of this close link between learning outcome and testing procedure is supposed to lie in the clarity and precision thus afforded the teaching, learning, and evaluating processes. The teacher will know exactly what to teach, the student will know what to learn, and the evaluator will know how to test. The disadvantage, however, comes from the more and more atomistic approach taken in specifying learning outcomes. In an effort to secure the requisite agreement in judgment, the content of the subject and the behaviors associated with learning the subject are analyzed into smaller and smaller and more and more standard learning outcomes. Since the evaluation techniques are tied precisely to the learning outcomes, the test, too, will emphasize the fragmented and the standard as opposed to the integrated and the innovative. Every effort to avoid such a routinization will be a potential threat to the agreement in judgment which must underlie the objective (unbiased) nature of the test. It is a no-win situation.

Another way of making the same point is to note the extreme temptations to "teach to the test" in such a situation. The more fervently one believes that the specific intended learning outcomes should be logically tied to the test, the less outrageous will teaching to the test seem. After all, those specific test behaviors are the goals of the course. Indeed, Stake (1973, p. 207) has noted that the performance contractor in the Texarkana project defended himself against charges of teaching for the test by noting that "teaching and testing had been directed toward the same specific goals, as should be the case in a good performance contract." Teaching to the test becomes virtually a logical injunction!

There is a second educational effect of conventional testing wisdom. The implicit model of testing for learning seems to be a sampling from a field of predetermined possible responses. The total field is presumably determined in principle by the subject matter; one supposes that the student has acquired some, if not all, of these responses, and the tester attempts to sample the student's response repertoire to see how closely it matches the total field. This picture of the situation gives rise to a whole host of puzzles regarding determination of "the" field of study, what constitutes a representative sample, how to sample a student's repertoire, and so on.

The major difficulty, however, seems to me to lie in the implications concerning how what has been learned gets applied to new situations—the so-called transfer of training question. How does a student recognize a new situation as one in which old principles are appropriate? Gronlund, for example (1976, p. 33), admits that an infinite number of possible learning outcomes can be specified for each general objective. This is, presumably, because new situations can vary indefinitely. There are logical questions of how to specify an infinite set without making use of precisely those general goals one is trying to specify, but let me waive *that* objection.

Somehow, the student must recognize the test question as sufficiently like what has been learned to call forth a response. Next the student must carefully analyze the new, transfer situation determining in what respects it is like, and in what respects unlike, what has gone before. Then the appropriate response from the infinitude available must be chosen and applied to the new situation. Of course, we are typically wholly unaware of such delicate analysis of situation, choice of response, and application to new context. All of that would have to occur unconsciously. The fact to be explained is that we can and do use what we have learned in an indefinite number of new situations. We are incredibly adaptable. The question is whether the picture of selecting from an infinity of preexisting responses the one which just matches the given situation adequately accounts for the fact of transfer. Surely, we might at least consider alternatives.

Finally, the conventional wisdom on tests leads one to a picture of the repertoire of learnings being sampled as consisting of stable, static elements—facts, concepts, methodologies, skills, understandings, and so on. Gronlund again (1976, p. 29) distinguishes sharply between the process of learning—the educational experiences undergone—and the products of learning—the knowledge and skills achieved. It is the products which must be utilized in framing learning outcomes. It is the products for which we must test. It makes little difference if we have covered certain material if nothing has happened to change our students. So in evaluation we must look to see what changes have occurred and not simply list the educational processes carried out.

Now there is nothing really wrong with such advice. The old educational saw that there can be no real teaching without learning reminds us that the object of our instructional and learning processes must in the end be changes in the students. Too much education has gone on with a simple reference to what the teacher has done—never mind what the pupil has accomplished.

However, the pernicious outcome of this emphasis on outcomes is that we may be tempted to view the products of learning as static, relatively stable atoms, in short, as "things"—knowledge, skills, understandings, and the like. However, such a view is not a logically necessary result of recognizing the process-product distinction in learning. The product of learning may well be that students now have new ways of processing their experience. The product of a process may be another process. Perhaps we should speak not of knowledge, but of the processes of knowing (see Campbell, 1959, 1974), not of skills, but of activities, not of understandings, but of the process of understanding.

I am not urging that we concentrate on the *process* of learning itself, but rather, that we consider the possibility that the *product* of learning may well be processes, rather than substances. As Stephen Toulmin says in the epigram to his provocative book, *Human Understanding* (Toulmin, 1972, p. x) "A man demonstrates his rationality not by a commitment to fixed ideas, stereotyped procedures, or immutable concepts, but by the manner in which, and the occasions on which, he changes those ideas, procedures, and concepts."

There are two reasons which to me have emerged from the discussion thus far that counsel toward considering knowledge processes rather than knowledge products as basic to the evaluative situation. First, the agreements in judgment which seem to underlie our ability to be objective in any field of human activity require physical and mental *action* in the world and not mere copying of the world. In a sense, we must contribute to the construction of our knowledge. This theme is common from the philosopher, Kant, through the psychologist, Piaget, to the devotees of open education. Second, this construction seems to take place not through the application of set procedures, but through the activities of seeing things as belonging to certain contexts. One and the same "thing" (the circle) can be seen as olive or navel depending on the context of activities in which we might place it. One and the same thing (a story in an elementary reader) can be seen as exemplifying family values or as sexist depending on the context in which we read the story.

The cybernetic alternative

I contend that a cybernetic approach to testing for learning gives promise of overcoming the problems I have uncovered with the conventional view of what testing for learning is all about. The cybernetic view does, I believe, take knowledge as basically a process rather than a substance. In a straightforward way a cybernetic view shows how our knowledge is constructed through our activities and how feedback loops provide the context for interpreting specific examples of action. More to the point for education, cybernetics shows how the novel and innovative response in an evaluative setting can be just as objective as the standard and paradigmatic. The phenomenon of transfer of training becomes transparent in a cybernetic setting, for within its effective range of control, what feedback loops do is control for the kinds of variations found in typical transfer tasks. Finally, the puzzles of how the evaluator samples from an infinite repertoire disappear. On the cybernetic view, one does not have a set of atomistic responses from which to choose, but rather a control system whose operation is designed to counteract disturbances.

What I will do in the remainder of this chapter is sketch the basic model of the testing process from a cybernetic perspective. I will try to show how the difficulties I raised above for the conventional view can be accommodated within the cybernetic model and how the subjective-objective distinction among types of tests with which I began this paper collapses when viewed from the cybernetic perspective.

I shall not spend the time outlining the cybernetic perspective itself but shall assume the model described with great clarity by William Powers in his *Behavior: The Control of Perception* (Powers, 1973) and summarized by him in this volume.* My thesis is simple. The cybernetic or control system approach gives promise of overcoming the accumulated conceptual and empirical difficulties of the conventional approaches to testing for learning and at the same time points to fruitful new directions for expanding our conceptions of what objective (unbiased) testing must be like. Finally the cybernetic approach accepts in the main the current common-sense forms of testing and merely reinterprets what we are doing when we engage in standard testing practice. No new model such as the one I am proposing can ignore what has proven successful and fruitful under past paradigms. What it can do is to point to new directions to proceed from here. With that methodological preamble, let me begin.

I have described the conventionally accepted model of testing for learning as sampling from an hypothesized infinite repertoire of learned atomistic responses. The cybernetic model of testing for learning can perhaps best be described as introducing a disturbance to an hypothesized knowledge process and observing to see if the disturbance is counteracted, just as a thermostat counteracts the disturbance of falling temperature by turning on the furnace. Basically, I am suggesting that for each of the typical sorts of things we learn and come to know, a control system (or system of control systems) can be hypothesized constituting that knowledge. Coming to know is coming to have the appropriate control system. Knowing is operating with the control system or, in the dispositional mode, is being ready to operate with the control system if a disturbance occurs.

I want to stress that what a control system *does*, as Powers so beautifully illustrates in the title of his hook, is to control perceptions by behavior. Control systems do *not* delicately vary their outputs to match the details of varying situations. Rather, when they sense a disturbance in the quantity they control, they drive their outputs relatively blindly.

---

* "This volume" refers to the volume where this chapter appeared. Powers' summary: *Cybernetic theory: Research models in child development* is reprinted in Powers, William T. (1989). *Living Control Systems: Selected Papers of William T. Powers*. Bloomfield, NJ: Benchmark Publications.

If the system is at all well adapted to typical ecologies in which it finds itself, the outputs will operate through the environment to counteract the disturbance. The control system wants to "perceive" a certain state and will operate to the limit of its capacity in attempting to perceive that state. Even the lowly thermostat exemplifies this crucial feature. *What* the thermostat does is to control the ambient temperature, and it operates the furnace in case the sensed temperature differs enough from the set temperature. And it will do so whether or not the house is well, or poorly, insulated, whether or not there is an open window or a lamp near the thermostat, and whether or not there is a burner clogged on the furnace. The system will simply drive its output if it senses a difference (or disturbance) between what is actually perceived and what the system wants to perceive. Even though a *causal* analysis of the system may go from input through the comparison of input with reference signal to the output and thence through the system's environment back to the input, the *explanatory* direction is just the opposite. The line of control runs through the perceptual side of the system (Powers, 1973; Petrie, 1974; Ryan, 1970). *What* the system does is control its inputs or perceptions.

How might this work in the case of testing for learning? Consider first the knowledge of facts. On the cybernetic model, knowing a fact is being in possession of a feedback system which represents that fact as a reference signal in the system. The input function of the cybernetic system monitors the environment for the degree to which it is like the given fact, and if that perception does not match the reference signal, an output will occur that can change the environment to bring perception and reference signal into line. For example, if someone knows that plants need sunshine to grow and perceives a plant in the dark, he'll move it, conclude the plant won't grow, or do something else which removes the disturbance.

How does one test for knowledge of facts? The technique is to devise some perceptual situation such that if the student being tested knows the fact, the situation will be perceived as containing a disturbance to the hypothesized cybernetic knowledge process that represents the fact. All the evaluator has to do is observe whether the student's output tends to counteract the disturbance or not. If it does, then the fact is probably known; if not, the student probably doesn't know the fact. The approach is not fool-proof. The disturbance may be removed fortuitously (e.g., the student may guess when the fact is not known), or the student may fail to remove the disturbance even though the fact is known (e.g., the test question may be misunderstood). But this is no more a limitation on the cybernetic than it is on the conventional approach.

Of course, most test situations are constructed to guarantee the motivation of the student and the likelihood that the control system representing the fact will be engaged if the student knows the fact. Thus, the disturbance might be a true-false test with the item being "The Declaration of Independence was signed in 1776." In this case, marking "false" would constitute a disturbance and if the student knows the fact, the disturbance will be avoided rather than directly counteracted.

What is the situation with regard to the learning of concepts? Once again the model suggests that a concept is represented by a cybernetic control system. Suppose the task is to learn the concept of "goose" where geese have to be distinguished from ducks and swans (see Kuhn, 1974). In this case, if one has learned the concept of "goose" then perceptual examples of geese will be recognized as such. A swan would constitute a disturbance to the "goose" control system possibly because of the swan's arching neck. The disturbance would be removed by classifying it as "non-goose." A duck might disturb the goose control system because of its shorter neck. Again the disturbance could be removed by excluding the instance from the concept.

How does one test for knowledge of concepts? He can introduce disturbances and see if they are counteracted or give samples of positive and negative instances of geese and see whether the disturbances are corrected or avoided. A multiple choice question could also be used, as,

A goose is a
a) bird
b) fish
c) mammal
d) none of the above.

Notice the close connection between concepts and propositions on the control system model. Recognizing certain perceptual experiences as instances of concepts seems to involve making judgments of the sort, "This is a goose." At the same time instances of concepts stand in many relations to other items in the environment. So concepts figure in many propositions, many of which would, if false, constitute disturbances to controlling for the concept itself. Thus if an object were not even a bird, it could not be a goose.

On the other hand, it used to be thought that whiteness was always associated with swans. The black swans of Australia did not constitute a large enough disturbance, however, to render those birds non-swans.

In this way a rough and ready categorization of essential and accidental properties of any concept can be worked out. Notice, however, that the categorization will be relative to an individual's system of control systems. Does the absence of a given property cause a large enough disturbance to remove the example from the effective range of control of the hypothesized control system (e.g., this object is not a bird)? If so, the property is essential. If the property's absence leads rather to corrective behavior (e.g., the acceptance of black swans), then the property is accidental *for the given system* of control systems.

How does understanding fit into the control system model? Items of varying complexity can be understood—concepts, facts, people, theories, etc. Each of these needs to be analyzed to determine just what is at stake. Nevertheless, some general comments are probably possible. Understanding generally involves placing something in a larger context (Halstead, 1975). In control system terms, this means that when someone understands something, there is a control system nested within a larger complex (probably hierarchical) of control systems. Thus if one not only knows that plants need sunshine to grow, but also understands why, it means that some knowledge of the theory of photosynthesis is also part of the student's control systems and further that these control systems subsume the lower-order facts of growth in sunshine.

How does one test for understanding on the cybernetic model? Because of the way larger contexts provide ways of assimilating whole classes of experience, testing for understanding is less susceptible to "objective" tests than testing for factual knowledge. Disturbances to understanding can occur in a large number of connected ways, as can the corrections to those disturbances. One could demonstrate an understanding of photosynthesis by articulating the theory, by designing appropriate "growth" lights to take the place of sunshine, by conducting appropriate experiments on photosynthesis, and so on and on.

On the conventional view of testing for learning, testing for understanding tends to move one into the area of "subjective" tests. But this can now be seen as a highly misleading way of putting the point. On the cybernetic view, the test is always to introduce a disturbance and see if it is corrected. Thus tests for understanding are just as objective (unbiased) as are tests for knowledge of facts. The difference is that the available area for introducing disturbances to control systems representing understanding is so much larger and so many more things will count as corrections to the disturbance.

Consider the analogy with the thermostat. It controls temperature alone. Suppose we add a humidistat to control humidity and tie both the thermostat and humidistat together with a "comfortstat" that controls a complicated combination of temperature and humidity, so that a whole range of varying combinations of temperature and humidity could satisfy the comfortstat. The comfortstat represents an analogue of "understanding" temperature and humidity relations. The ways of disturbing this system are vastly increased over the ways of disturbing the thermostat alone (analogous to a single fact). Similarly, the ways of correcting disturbances are much greater. We can change temperature, or humidity, or various combinations of them. Imagine what would happen as we increased the interrelatedness (understanding potential) of such a system. Clearly the attempt to specify in advance the variety of specific potential disturbances and corrections becomes impossible. Rather, the global feature of "comfort" becomes the explanatory concept.

Another item sometimes said to be taught, learned, and tested for is methodology. This, too, covers the waterfront just as does "understanding." Different fields have different methodologies, and there are a host of degrees of explicitness of methodology from formal algorithms to less precise heuristics to "the scientific method" which may itself change over time. Once more the cybernetic view treats methodologies as represented by control systems (or systems of control systems). The methodology for adding a column of figures is fairly precise and the activities performed in that process are easily checked, both at the time of performance and subsequently. At the more general level of "good science," where physicists are, for example, debating the fruitfulness of searching for ever more fundamental particles, "disturbances" and "corrections" will be much more idiosyncratically determined.

How does one test for a given methodology on the cybernetic view? Once again, see how disturbances are corrected. With something as explicit as the rules for addition, the task is easy. Give problems and see how the student avoids disturbances. Be sure to throw in some "hard" ones, too, in which the disturbances are difficult to catch. For less precise methods, such as heuristics, again introduce disturbances and see how they are corrected. Various chess moves and situations can be seen as disturbances to the heuristics of how to play good chess. So far, chess masters seem to grasp the heuristics of chess much better than do chess-playing computer programs. Is it because chess programs must analyze the heuristics into digital processes, whereas human heuristics work on an analogical gestaltist basis? I tend to think so. At the level of "good physics," only history ultimately can judge whether a proposed "correction" really is one. Did it, in the end, work out?

A particularly interesting class of items taught and learned is commonly called skills. Consider the ability to drive a car, for example. What does such an ability look like on the cybernetic model? Take the simple aspect of steering. Most drivers, if they think about it, will say that they steer by maintaining a certain perceptual relationship, e.g., keeping the hood ornament on the right hand edge of the lane in which one is driving. Indeed grasping such a relationship is a great advance in learning to drive. One no longer has to analyze the situation and decide what to do next at each moment in the driving sequence. Indeed, when one does analyze and "think about" one's driving by "applying" the principles, overreaction almost invariably ensues. The good driver does not say, "Aha, a curve to the right of about 30 degrees, so, given the steering ratio of my car, I must turn the steering wheel to the right about 10 degrees." Rather the driver notices a disturbance to the desired perceptual pattern and acts until the ornament gets back on the edge of the lane. The driver controls perceptions, not outputs.

Notice, too, how much easier the cybernetic view makes the account of handling situations which are never analyzed except insofar as they disturb the desired perceptual relation of ornament to edge of road. Consider a crosswind. The driver need never (and often does not) recognize there is a crosswind. But insofar as the wind affects the position of the ornament relative to the edge of the road, the driver will automatically correct the disturbance. Similar remarks apply to road conditions, looseness in the steering mechanism, and so on. The driver controls perceptions, not outputs.

How do we test for such skills? In the driving case, it's easy. There are enough natural disturbances in driving so that we put an examiner in the seat beside the driver and let him observe how the driver counteracts the disturbances.

One can also easily see the difference between skills which may be possessed and tendencies or habits which we hope will be exercised. The *ability* to keep household accounts may be checked by classroom tests; the *tendency* to do so would require what might be considered an unwarranted invasion of privacy. On the cybernetic view, the difference can be seen as a difference in *what* is being disturbed, the control system which wants to get a good grade in a school subject or the control system for keeping track of day-to-day expenses (Petrie, 1974).

In either case, the cybernetic model of testing for performance skills seems *prima facie* much more realistic than the conventional model. We do not learn the principles of skills and then apply them. Rather we acquire certain perceptual quantities to control and correct for disturbances. One could not possibly specify all the behaviors that go into correctly steering a car and then go about testing for each of them. Rather we know perfectly well what steering a car properly means, and can see when someone does it even if the disturbance is a blowout or a child darting into the street after a loose ball.

A more complicated example

None of the examples of tests one might use on the cybernetic model has differed much, if at all, from the kinds of tests one might use on the conventional model. I have suggested multiple-choice, true-false, identification, skill demonstration, and the like; standard testing uses these types too, and it might be objected that all I have done is to introduce a jargon-loaded way of talking about these familiar kinds of tests.

I have already admitted that any new theoretical model must handle the standard examples in the field. Any proposed new theory that said in effect that everything we've been doing has been wrong would, for that reason, be highly suspect, so the cybernetic view must accept most standard testing procedures. It is in the borderline cases and in giving us hints on how to extend our understanding that theories will differ. In this section I want to try to sketch a borderline case where conceptualizing the testing situation under the conventional versus the cybernetic models might lead to different results.

Consider the learning of fractions. This task is typically a difficult one for many students. One of the more difficult aspects is getting the student to grasp the idea that a whole can be cut up into parts (fractions) and that these fractions make up the whole. A typical test item at the beginning stages of teaching fractions might go as follows: Sue, Bill, and Mary went to a pizza parlor to order a large pizza. If they divided it equally, how much did each one receive? The answer is, of course, one-third. This kind of test item is probably repeated in slightly different formats dozens of times during a student's learning fractions. One varies the food and the number of people, but the test item remains basically the same.

And yet, as teachers well know, students can become fairly proficient on this type of item and yet not fully grasp the idea that those three thirds make up the whole pizza. On the conventional model, one simply searches for other preexisting responses to situations which call for different and perhaps more complicated applications of the rule of dividing the food by the number of people. One perhaps also hopes that the division of food will generalize to the division of toys, money, jobs, or whatever. But the basic concept is that the rule is applied in new situations.

What might a test item look like on the cybernetic model of introducing a disturbance to an hypothesized control system? Suppose the concept being tested for is the part-whole relationship. Consider the following item: Sue, Bill, and Mary went to a pizza parlor and ordered a large pizza. Sue said, "I want one-third of the pizza"; Bill said, "I want one-third of the pizza"; and, Mary said, "I want two-thirds of the pizza." Can each child receive what he or she wants?

In this item a disturbance to the whole-part relationship has been introduced and how the child corrects the disturbance will show whether or not the concept has been learned. The item can be varied in a number of ways and leads itself easily to follow-up questions. For example, one might add, "If so, will there be any left? How much?" "If not, why not?"

The item can then be varied by having each of the children ask for one-fourth of the pizza. Then with the follow-up questions, a set of parts which do not add up to the whole can be identified. This is another, different, disturbance to the part-whole concept for which we are testing. The point is not that the cybernetic view generates tests or test items which could not in principle be generated under the conventional model, but rather that it sets us off in a slightly different, and, I think, more fruitful direction.

The reaction to this kind of item by the advocates of the conventional model of testing will be that they too could have come up with the item. It requires just a complicated bit of transfer of training. It requires the student to select from a broader response repertoire. And yet such a reply doesn't quite ring true. Intuitively one is inclined to say that the student hasn't really understood the part-whole relationship until he or she can answer such test items. It's not that the test requires a new application; it's rather that the test gets at the concept directly.

If this sort of example is at all persuasive, I suggest it is because it makes direct use of the model of removing a disturbance rather than the model of selecting from a repertoire of responses. However, I realize no single example can possibly make my case. Rather it is the general theory or model with which one approaches each case that makes the difference. For that reason, I want to consider in the next section how the cybernetic model deals with the general problems associated with the conventional model.

Cybernetics and innovation, transfer of training, and knowledge processes

Recall the three main difficulties I uncovered with the conventional model of testing for learning. These were, first, that the conventional model finds it tremendously difficult to find room for appropriate, yet innovative, test responses. The push for standardization and stereotyping of response is strong. Second, the conventional view with its pictures of sampling from a predetermined repertoire of atomistic responses renders transfer of training quite problematic. The idea that somehow we (unconsciously) analyze each new situation and then select just the right response from the repertoire seems dubious at best and simply a redescription of the problem at worst. Third, the conventional view pictures a relatively stable product as the result of the learning process, and it is that product that is sampled during testing. I have urged that there is strong reason for considering knowledge itself to be a process rather than a structure. Thus if we are testing for knowledge, we are testing for certain kinds of processes.

How does the cybernetic model of testing for learning fare with these difficulties? The picture of introducing a disturbance to a hypothesized control system and checking to see if it is counteracted or avoided handles the novel response problem in a most illuminating way. Not only does the cybernetic model wait to see what the response is before deciding whether it's appropriate, it also gives a method for determining whether the response is appropriate. Does it tend to remove the disturbance? Thus, the model does have empirical consequences and the existence of hypothesized control systems can be tested. The very mark of human rationality seems to be its adaptability to an indefinite variety of circumstances. In short, the *novel* answer to "test" situations is the core of our concept of knowledge. Instead of its being an unwelcome intruder for whom we must somehow find room in the conventional model of testing, the novel answer is the master of the house on the cybernetic view. By giving pride of place conceptually to this central feature of human adaptability, the cybernetic model is in a very clear sense a more human approach to testing for learning than the conventional approach. How could we ever have thought that we could select from among an infinity of potential responses anyway? Computers may check off lists; humans seldom do.

The case with transfer of training is very similar. Transfer of training is a problem on the conventional view but becomes a paradigmatic feature of the cybernetic view. On the conventional view it is hard to see how one recognizes the new situation as sufficiently like the paradigm situation in which the knowledge was acquired to call forth an appropriately adjusted response. One must "apply" knowledge to new situations and the temptation is great to assume that another piece of knowledge is present that tells us when and how to apply our original knowledge. But in this way lies a dark and infinite regress. For we need also to know when and how to apply the rules of application, and off we go.

But the regress generated by viewing the transfer problem as one in which we must "apply" knowledge simply disappears on the cybernetic view. For on the cybernetic view there is not the slightest temptation to speak of applying knowledge in new situations. On the contrary, the situation is perceived as more or less like the paradigmatic learned situation, and if the control system is operating, the student will simply behave in ways that reduce the "distance" between the perceived and the desired situation. The thermostat causes the furnace to go on, which reduces the difference between the perceived ambient temperature and the thermostat setting. The car wheel is turned to reduce the perceived distance between the hood ornament and the edge of the road. The physics student learns to see the problems at the end of the chapter in terms of the principles learned in the chapter (Petrie, 1976; Kuhn 1974). In short, a cybernetic system operates to change the environment to bring it closer to the organizing principle or reference signal of the system. One does not "apply" knowledge at all, and so there is no problem of how application can occur in new situations.

Finally, the fact that a control system is operative in the production of the quantities it controls clearly indicates that one is dealing with a knowledge process and not with a static structure of knowledge. Control systems exhibit on their face the contribution of the system to the construction of knowledge. Notice that it is only a *contribution* to knowledge and not the whole constitution of it. If a thermostat is hooked up to an inefficient furnace in a poorly insulated house during very cold weather, it may be unable to stabilize the ambient temperature at 68° F, although it may tend in that direction. Similarly we cannot all, simply by wishing it so, run a four-minute mile or win a Nobel prize in physics. Cybernetics shows how control systems perceive in the environment whatever it is they are controlling for and how they can change the environment *within limits* to become more like what is being controlled for. The limits are set by the world as it is, natural and social.

Objective and subjective tests reconsidered

To conclude, I shall return to the distinction between objective and subjective tests with which I began this paper. In a very real sense the distinction between mechanical and judgmental scoring procedures no longer seems very important. Indeed, all evaluative measures depend at base upon agreement in judgment, and so, if anything, "subjective" tests are more central to evaluation than "objective" tests. But the question then arose as to what the agreement in judgments was agreement about—a static repertoire of atomistic responses, or the correction or avoidance of disturbances to control systems. I have urged that the cybernetic approach gives promise of avoiding some of the problems with the conventional approach.

No new theory of testing would have much plausibility if it urged a wholesale rejection of the kinds of testing currently in use. It is highly unlikely that our best professional practice, accumulated over the years by a trial and error process, is largely wrong. It wouldn't have survived were that the case. Nevertheless, when it comes to understanding our common practice and extending it in new directions, different theories do make a difference. On a global scale this is reflected by the kind of second-class citizenship enjoyed by "subjective" tests under the conventional theory of testing. On the conventional view we must justify using "subjective" tests because of our inability to find "objective" ones. On the cybernetic view the burden of proof is reversed. The basic testing situation is seeing if disturbances are corrected *in whatever way the student sees fit*. The novel response rather than the typical one is the standard condition. On the cybernetic view we must justify using objective tests that limit the testing situation so that only a few responses will count as "right." No doubt such a justification can be given in certain cases, but on the cybernetic view, it *must* be given.

Gronlund (1976, p. 149) has interesting comments on the effect "objective and subjective" tests have on the control of student responses. He admits that an objective test "limits pupil to type of response called for." However, he seems to view this as a virtue because he continues that this "prevents bluffing and avoids influence of writing skill." His comments on the essay test concerning its control of pupil response are that "freedom to respond in own words enables bluffing and writing skill to influence the score." Clearly, the cybernetic approach would render almost the opposite judgment on the value of structuring the pupil's responses. If *what* is being done in testing is seeing if and how disturbances are counteracted, then limiting the ways of counteracting the disturbances is *prima facie* wrong and can be justified only in special circumstances.

Multiple-choice tests, for example, will continue to be used about as much as they are now. But in the cybernetic view they are seen for what they are, an often artificial limiting so that only five responses seem appropriate for removing disturbances. It helps, of course, to tell the student to choose the "best" response, but the point is that the ways of removing disturbances have been set by the tester. As is granted even under the conventional view, multiple-choice tests are best in content areas where the facts are pretty well agreed. When used to test for more complex learning such as understanding or reasoning, multiple choice exams are only plausible if the only appropriate responses called for are unitary and sterotypical. If understanding in an area is complex and unpredictable, the use of multiple-choice exams will distort the subject matter. Thus, using multiple-choice exams in mathematical reasoning is probably all right, but their use in testing for the interpretation of a poem would be misleading since the interpretations have been limited.

The true-false test is even more constricting than the multiple-choice exam. It presupposes a clear binary decision about a given item. Again this may be true for some limited fields and for certain limited areas of those fields, but the true-false exam begins to mislead when it is extended beyond obviously binary decisions. I shouldn't be surprised if the barbarisms "more true," "less false," and the like are not somehow linked to trying to extend true-false items beyond their appropriate range.

Matching tests are useful for checking fine discriminations. If a host of similar items in one list "go with" certain items in another list, making sure these lists are matched is a useful way of checking our facts. A typical example is to match explorers with countries in a social studies test. Once again, the cybernetic view can account for successful performance in an absolutely straightforward way. Putting the country with the proper explorer literally removes the disturbance caused by having the column of explorers not match the column of countries. However, once more the subject areas in which such tests can be justified are probably a good deal fewer than are commonly believed. And the burden of proof is on those who would expand matching tests beyond a limited arena.

Completion or short answer essay tests are an interesting category. Viewed conventionally, they are probably closer to "objective" tests in that a fairly definite answer is usually sought by the examiner. Nevertheless, the student can respond as he or she sees fit. A wrong answer on the conventional view is just a wrong choice of response. On the cybernetic view, however, the student believes that the answer given will remove the disturbance caused by the blank in the completion exercise. This almost forces the teacher to ask, "What control system *was* operating for the student?" Valuable clues are available for diagnosing the student's mistake. What would the student have to believe for the wrong answer to appear right? Of course, good teachers have always used wrong answers diagnostically. It's just that the theory of selecting from a repertoire of responses gives us no guidance as to how to diagnose the error. The cybernetic view, on the other hand, asks the evaluator to view the response as attempted correction: in virtue of what control system would it be a correction?

One also needs to consider the test response *both* from the point of view of what the student believes is being controlled and from the point of view of the collective understanding the teacher is testing for. In short, there are two cybernetic systems in operation—the student's and the teacher's. Do they control the same quantity? Essay tests bring out this duality very nicely. Because of the relative lack of prespecified structure in an essay exam, the student can remove the disturbance in whatever way seems most appropriate to the student.

The teacher-evaluator probably has a *better* view of what control system the student possesses from an essay test than from a multiple choice or completion exam. In the essay test the student demonstrates in a fairly complex way how his or her cybernetic control systems are organized. Literally, more of the "person" can be discerned through an essay test than through "objective" tests. Concomitantly, the teacher-evaluator can judge the extent to which the student's control systems match up with the teacher's (assuming the teacher embodies the collective disciplinary wisdom being taught). In this way the phenomenon of teachers learning something from their students on essay tests is explained by the fact that the teacher must decide whether the student's effort causes any disturbance in the teacher's control systems.

Can the student "bluff" more easily on an essay test? In a sense, he obviously can. But the bluffing will only be effective on less than competent teachers whose *own* control systems do not enable them to distinguish the glib from the insightful. "Objective" tests with their mechanical scoring procedures do tend to protect student assessment from relatively incompetent teachers, but at the cost of limiting the more competent students. It's not clear that such a trade-off is always in the best interests of education.

Intuitively, university professors recognize the value of the oral exam as used, for example, in the defense of a doctoral dissertation. The oral exam, far from going off on tangents and not "covering the ground," enables the evaluators to explore in some detail the student's cognitive functioning. The professor asks a question introducing a disturbance into the student's control systems. The student responds in a way he believes will remove the disturbance. But does it from the evaluator's point of view? Well, if the student really has "got it," and responds in the way he or she did, then this further question should elicit such and such a correcting response. And the process can be iterated again and again. The depth and complexity of modes of correcting disturbances that can be explored in an oral exam is truly amazing. Surface answers will not do. One can "cram" for an "objective" test, one cannot "cram" for an oral. The former gets at breadth, but probably in a superficial way. The latter gets at the enduring knowledge processes of the student.

The much maligned interview also turns out to be a very potent testing device on the cybernetic view. Its advantages are very similar to those of the oral doctoral exam. The ways in which disturbances are corrected by the interviewee tell a great deal about the knowledge processes likely to be employed later on. How does the candidate react to a change of pace? Does the person give short or long answers? What specific kinds of things does the interviewee know? If one were sampling from a repertoire of atomistic responses, the interview would probably be incredibly inefficient and subjective. But if one is looking to see if and how disturbances to control systems are corrected, the interview can pursue these in great depth.

One can even test by observation, particularly by participant-observation. I suspect that most day-to-day evaluations by classroom teachers are of this type. Classroom teachers, especially in the elementary grades, interact with their students in a wide variety of contexts. They really can get to know their students, and they are constantly introducing disturbances, observing how these are corrected, and devising remedial work. The difference between observer and participant-observer on the cybernetic view is the difference between having to wait to see how natural disturbances are corrected and being able to introduce those disturbances. The classroom teacher is probably a paradigm case of the participant-observer category of evaluator.

One of the traditional objections to the observer, or even participant-observer, methodologies is that the observations tend to be subjective, i.e., couched in the observer's categories. This objection is easily met on the cybernetic view. Any hypothesis about what the student knows will be cast in the observer's categories. However, there is a check for the adequacy of that characterization. Introduce what would be a disturbance if the hypothesized control system is operative in the student and see if it is counteracted. Of course, mistakes are still possible. An hypothesized system may be close enough to the actual system to be judged as present on the basis of the disturbances introduced when it actually is not. But this possibility of error always obtains in empirical sciences. It is sometimes very difficult, although not in principle impossible, to disentangle competing hypotheses. The cybernetic view does indicate the general principles to be followed in disentangling the hypotheses.

Another objection to observational methods has been that sometimes what people really know and believe is not what they say. So even if one asks a student how *he* views the situation, that may not reflect his "real" cognitive state. The cybernetic approach seems to me to obviate that criticism by treating the correction of disturbances in a much broader context than simply what people say. The cybernetic approach concentrates on what people do. What people say is part of what they do, but only part. The general formula of introducing a disturbance and seeing how it is counteracted draws attention to the whole field of action, not just verbal actions,

The use of rating scales and check lists in an effort to make the observation more "objective" will probably only end up in blinding the observer to other important features that do seem to be making a difference. Of course, the other side of that coin is that checklists can perhaps help the inexperienced observer keep track of the important things to be watched for.

The "case-study" evaluation utilizing observational techniques in a central role is a perfectly valid mode of evaluation. When competently undertaken, it will reflect what the actors know and do. In brief, it will illuminate what is going on in the sense that this is what the actors are actually doing. For on the cybernetic view the evaluation introduces a disturbance to an hypothesized control system and sees if it is corrected. Case studies can be generalized precisely to the extent that similar people in similar situations tend to *do* the same things. Notice carefully that what someone *does* is defined not in terms of behavioral effects—those could well vary from situation to situation—but rather in terms of the quantities the actor is controlling. Thus, providing remedial classes in one context, e.g., a community college, may be just as much the teaching of good writing as refusing to provide such classes in a different context, e.g., a selective private college. The level of analysis of what is going on must always be in terms of what disturbances will be corrected.

In summary, the cybernetic model of testing for learning promises a number of advantages over the conventional model. The conventional picture of sampling from an infinite set of discrete responses seems to generate problems of how to evaluate innovative responses and how to understand transfers of training. It also presupposes the somewhat dubious idea of static knowledge structures. The cybernetic view of observing to see if and how disturbances are removed seems to solve all of these problems at a stroke. The common wisdom of dividing tests into subjective and objective is misguided. All evaluation depends on agreement in judgments among those who know what the activity being evaluated is. "Subjective" tests turn out to be closer to what students actually know and can do and are, and therefore are more human (not merely more humane) in an absolutely straightforward sense.

In changing from the conventional to the cybernetic view, the burden of proof is shifted radically. No longer must one justify essay tests because really objective tests are not available. On the contrary, one must justify using objective tests by trying to show that the stereotypical responses encouraged are really all right in the context. Finally, I am not claiming that one must do away with the commonly accepted and used testing procedures. Any new theory must account for what we have already been doing well. The difference the cybernetic approach will make is on the borders and in the directions in which testing for learning might expand. And *that* may make *all* the difference.

References

Campbell, Donald T. 1974. Evolutionary Epistemology. In *The Philosophy of Karl Popper*, Vol. 14, 1 and 2, *The Library of Living Philosophers*, P. A. Schilpp, ed. LaSalle, Ill.: Open Court.

Campbell, Donald T. 1959. Methodological Suggestions from a Comparative Psychology of Knowledge Processes. *Inquiry* 2:152:182.

Gronlund, Norman E. 1976. *Measurement and Evaluation in Teaching*, 3rd ed. New York: Macmillan.

Halstead, Robert E. 1975. Teaching for Understanding. *Philosophy of Education.* Philosophy of Education Society, 52-62.

Kuhn, Thomas. 1974. Second Thoughts on Paradigms. In *The Structure of Scientific Theories*, Frederick Suppe, ed. Urbana, Ill.: University of Illinois Press.

Petrie, Hugh G. 1977. Comments on H. S. Broudy's "Types of Knowledge and Purposes of Education." In *Schooling and the Acquisition of Knowledge*, R. C. Anderson, R. J. Spiro, and W. E. Montague, eds. Hillsdale, N.J.: Lawrence Erlbaum Associates, 19-26.

Petrie, Hugh G. 1976. Metaphorical Models of Mastery: Or How to Learn to Do the Problems at the End of the Chapter in the Physics Textbook. In *PSA, 1974*, R. S. Cohen et al., eds. Dordrecht, Holland: D. Reidel, 301-312.

Petrie, Hugh G. 1974. Action, Perception, and Education. *Educational Theory* 24:33-45.

Petrie, Hugh G. 1971. Science and Metaphysics: A Wittgensteinian Interpretation. In *Essays on Wittgenstein*, E. Klemke, ed. Urbana, Ill.: University of Illinois Press, 106121.

Powers, William T. 1973. *Behavior: The Control of Perception.* Chicago: Aldine.

Ryan, Alan. 1970. *The Philosophy of the Social Sciences.* New York: Pantheon.

Scriven, Michael. 1972. Objectivity and Subjectivity in Educational Research. In *Philosophical Redirection of Educational Research: The Seventy-first Yearbook of the National Society for the Study of Education*, Lawrence G. Thomas, ed. Chicago: National Society for the Study of Education.

Smith, Ralph, ed. 1975. *Regaining Educational Leadership: Essays Critical of PBTE/CBTE.* New York: Wiley.

Stake, Robert E. 1973. Measuring What Learners Learn. In *School Evaluation: The Politics and Process*, Ernest R. House, ed. Berkeley, Calif.: McCutchan, 193-223.

Toulmin, Stephen. 1972. *Human Understanding, Volume I.* Princeton, N.J.: Princeton University Press.

[1985]
Testing for Critical Thinking

My thesis today is fairly straightforward. I hold that if we, as educators, do not provide a sensible epistemological basis for understanding the operation of the so-called "higher order cognitive skills," then current testing policies and practices will, in the name of critical thinking, promote almost its exact opposite.

I have divided my paper into four parts. First, I will attempt to isolate one key feature that seems to permeate much of the discussion and analysis of critical thinking. This feature is the element of monitoring, evaluating, and correcting errors in our thought processes. Second, I believe that the concept of adaptation captures this element of critical thinking in a most illuminating way. In particular, I will suggest that the operation of the goals, concepts, and standards which help structure experience and, therefore, partially define what it is to be an error, leads to the necessity of examining the processes of adaptation. Third, I will argue that current testing practice is ill-suited to focus on the multidimensional ways in which the process of adaptation can occur. Finally, I will explain how the consequences of failing to incorporate the notion of adaptation into testing practice may lead to a severely truncated and inadequate notion of critical thinking. Furthermore, because testing has become perhaps the major engine driving educational policy implementation, this situation could well result in some quite devastating educational consequences.

I. Critical thinking

An emphasis on higher order cognitive skills, especially what is called critical thinking, permeates much of the recent literature on educational reform. In the College Board's *Academic Preparation for College: What Students Need to Know and Be Able to Do*[1] under the competence of Reading, we find the

---

© Hugh G. Petrie. Presidential Address.
   Proceedings of the Philosophy of Education Society 1985. 3-20.

ability "to interpret a writer's meaning inferentially as well as literally."[2] Under Writing, "the ability to improve one's own writing by restructuring, correcting errors, and rewriting."[3] Under Speaking and Listening, "the ability to engage critically and constructively in the exchange of ideas."[4] Under Mathematics, "the ability to judge the reasonableness of a result."[5] Under Reasoning is located problem formulation and solution, inductive and deductive logic, recognition of fallacies, justifying conclusions, comprehension, and the ability to distinguish fact from opinion.[6] Similar emphasis can be found in many of the other national education reports and state action plans. Clearly, critical thinking is "in."

I wish to focus on an aspect of critical thinking that has been receiving increased attention in the psychological literature under the rubric of information processing theories of cognition. I refer to the feature of monitoring and evaluating thought. For example, Robert J. Sternberg in describing his componential theory of intelligence says:

> [M]etacomponents are the higher order or executive processes that we use to plan what we are going to do, monitor what we are doing, and evaluate what we have done. Deciding on a strategy for solving an arithmetic problem or organizing a term paper are examples of metacomponents at work.[7]

Monitoring and evaluating thought is certainly not all there is to critical thinking, but it is surely a key element.

Let us consider an everyday example of monitoring thought, or metacognition as it is sometimes called.

> It is a common experience while reading a passage to have our minds wander from the words. We see the words, but no meaning is being produced. Suddenly we realize that we are not concentrating and that we've lost contact with the meaning of the text. We recover by returning to the passage to find the place, matching it with the last thought we remember; once having found it, we read on with connectedness. This inner awareness and the strategy of recovery are components of metacognition.[8]

In monitoring reading the reader notices that the actual thought is not what is expected in terms of mental processing. There are purposes and goals to reading, usually implicit, but present nonetheless, as the detection of our mind's wandering indicates. Somehow we compare the actual situation with the goals and standards we have set ourselves and we can tell whether the actual situation does or does not come up to those standards.

Notice that this monitoring and evaluating is a very widespread phenomenon. For example, we might be considering making a certain step in constructing a logic proof and evaluate it on at least two different grounds. First, is it permitted by the rules of proof in the logical system? Second, even if permitted, is it likely to get us closer to the conclusion to be proved?

An example from physical activity might involve a basketball guard close to the basket being intimidated by Patrick Ewing. The guard notes the situation in light of the "goal" and modifies the behavior appropriately, putting more arc on the shot.

What is common to all of these examples of monitoring and evaluating is the presence of a concept, goal, or standard; a perception of the current situation; and an assessment of the "distance" between the current situation and the concept, goal or standard. Critical thought, although never *guaranteed* of success, involves a tendency to reduce the distance between the presently perceived situation and the concept, goal or standard, whether the thought be dispositional, propositional, or activity-guiding.

Thus in the case of our mind wandering while reading, we recognize that our daydreams do not match our goal for comprehension and we go back to where we last understood what we read. The step in the logic proof reduces the distance between where we are and the conclusion we wish to reach. In the basketball case, putting more arc on the shot eliminates the likelihood that Pat Ewing will put even more distance between the ball and the goal.

However, all of the examples thus far have involved activities which reduce the distance between the perceived situation and the goal in a quite specific way. They all *alter the current situation.* We pay attention to our reading, we make a justified step in the proof, and we put more arc on the ball. It is extremely important to notice that the distance between situation and concept, standard, or goal can logically be reduced by *changing the concept, standard, or goal as well as by changing the situation.*

Thus, in the reading example one could decide simply to skim the section for main ideas rather than aiming at total comprehension. In the basketball case, sometimes one should pass the ball back out rather than trying to shoot over an intimidating front line player. In the logic case, one could, and sometimes should, stop worrying about the proof and go to a movie to take a break from studying. Of course, simultaneous partial modification of any or all of the elements is also possible.

II. Epistemology

What kind of epistemology, then, is suggested by an emphasis on reducing the distance between perceived situation and concept, standard, or goal? In a paper they wrote for the National Commission on excellence,[9] Wagner and Sternberg "... characterize intelligence, when applied to the everyday world in which we live, as involving purposive adaptation to, shaping of, and selection of real-world environments relevant to one's life."[10]

Although Wagner and Sternberg use the concept of "intelligence," it is clear that they mean to encompass the monitoring and evaluating process I have been elucidating. Indeed, as noted above, Sternberg's theory of intelligence involves a metacomponential part whose primary function is to monitor and evaluate thought and action.

In *The Dilemma of Enquiry and Learning*, I argued extensively for an emphasis on knowledge processes, assimilation and accommodation, and the critical role of action in reaching a reflective equilibrium.[11] The epistemology sketched there was one of organisms adapting to the real world in which they find themselves. Like Wagner and Sternberg I also believe that an epistemology of adaptation is necessary for bringing the perceived situation into an equilibrium with our concepts, standards, and goals.

However, the reason that the equilibrium can occur is precisely because there is a tight conceptual link between the situation as structured or perceived and the concepts, standards, and goals in terms of which it is evaluated. This feature is of critical importance for understanding thinking in general and critical thinking in particular.

The point is Kant's and at one level is no more controversial than the fact that our concepts, standards, goals, purposes, and even the propositions we believe condition what we experience. Because position and momentum are important in physics we look at a body's location and movement and not at its color in applying principles of mechanics. Because color is important in crossing at a traffic light, we perceive it rather than, say, the swaying of the traffic signal in the breeze. Somewhat more controversially, those who view reading as decoding look only at pronunciation and fluency in deciding if someone can read. On the other hand, those who view reading as comprehending look to such things as literal, inferential, and evaluative comprehension and the extent to which students can learn other subjects *through* reading.

In other words, although our concepts, goals, and standards do not totally cause what we perceive, they are the conditions which, logically, structure our experience in one way rather than another. To this extent, at least, experience allows for the adaptation of thought and action to the world and cannot be "value-free."

In monitoring our thought and action, we usually try to change the world, i.e., the perceived situation. If the world proves recalcitrant, we can change our goals. If our goals are fairly firm, we often rationalize by bringing other goals and concepts to bear, thereby restructuring the situation so that it is no longer seen as problematic. Only occasionally do we alter our standards and even less often do we change our concepts in order to adapt. Furthermore, there are probably good evolutionary arguments for why these tendencies exist and are, in the main, justifiable. However true critical thinking can and sometimes should challenge basic concepts, standards, and firmly held goals. We can and should learn from experience.

Let me illustrate these somewhat abstract alternatives with a concrete example. At the University at Buffalo the faculty decided several years ago that all undergraduates should possess basic competency in mathematical skills. That was a goal or end decided upon. The concept of basic mathematical competency was then discussed and clarified. It came to mean competency in the kinds of mathematical skills and ideas covered in the Regents 11 (third year) high school mathematics syllabus. The application of this concept, however, also required the adoption of certain standards, i.e., a specification of the degree to which the competency must be present and the evidence that would so indicate. A passing mark on an entrance test was set as the appropriate standard. If students did not pass the exam, they could study the parts on which they were weak and retake the mastery test. Indeed, with typical university faculty arrogance it was determined that all entering students should take the test, even those who had passed Regents 11 math in high school.

The goal of guaranteeing basic mathematical competency for Buffalo students was to be achieved through acting in the world as I have described above. How, then, was the monitoring and evaluation of the necessary social thought and action to take place? What are the variety of ways in which critical thought concerning the mathematical competency of the students can occur?

As I have noted, the most common way for people to reduce the distance between concept or goal and the world is by changing the world. In this case, the whole design of the competency program is aimed at getting the students to demonstrate their mathematical competency either through obtaining initial passing scores or retaking the test after further study. The goal, concept, and standard were fixed and activities were undertaken to get students to achieve the goal.

What seems to be happening, however, is that a very high percentage of students are not passing the exam. We have, in other words, a distance between the perceived situation of actual competency and the goal of basic mathematical competency as determined by the exam standard. The monitoring process is telling us that the student activity is not achieving the goal. Of course, one can simply exhort the students to do better, but what other alternatives are available?

Consider the possibility of changing the *goal*. In real situations, one seldom simply gives up goals such as these, but since goals are usually the easiest things to change, we sometimes engage in a kind of goal *modification* which we tend to call rationalization. It has been suggested at UB that the goal of basic mathematical competency is really only appropriate for minimally prepared students. Obviously, the high schools are not doing their job; so that explains why so many students are failing. The goal is no longer mathematical competency for all; it is now competency for minimally prepared students. New concepts are brought to bear and the distance between the situation as now perceived—poorly prepared students—and the revised goal—basic mathematical competency for *college-level* students—is reduced. By changing the goal, an adaptation might be found.

Other approaches are possible. Consider the possibility of changing the *standards*. Some faculty have actually looked at the exams and suggested that they are really too difficult. Easier problems could be proposed. This would allow adaptation by changing the standards rather than changing the concept or goal. But that would "lower standards," not an easy thing for university faculty to do.

One might also adapt by changing *standards* in a different way. It has been proposed that incoming students simply take and pass Regents 11 math rather than an entrance exam. That would be initially adaptive except that it may be suicidal in the demographic situation facing colleges in the Northeast, i.e., lessening the problem in one area will create a problem in another area, as fewer students would qualify to attend the university.

Consider finally the possibility of changing the *concept* of mathematical competency. There actually is a group of faculty, who happen to be non-mathematicians, who are saying that, *those* questions on the test, even if made easier, are not really what they meant by basic mathematical competency in the first place. Furthermore, if these faculty are told that the questions are actually taken from Regents 11 exams, then they seem to be saying that their very concept of basic mathematical competency is changing. In short, the situation *could* even lead to a modification of the concept, although that is unlikely, given that such a change would open up these non-mathematicians to the charge that they really do not understand mathematics. After all, who are the keepers of the concepts of mathematics if not the mathematicians?

Thus, this concrete example shows quite nicely how the monitoring of thought and action can lead to adaptation in a wide variety of ways. Action in the world may be adaptive. Changing goals is sometimes adaptive. Altering standards may be adaptive. Even modifying concepts can be adaptive. Notice that I am making no claims at this point about what the best adaptation may be in any given case. I am simply pointing out that adaptation can occur in a variety of ways and that critical thought must allow for the variety.

The phenomenon of multiple modes of adaptation can even be discerned in the literature on the nature of critical thinking itself. For instance, in his classic *Harvard Educational Review* paper, Robert Ennis argues for a conception of critical thinking as the correct assessing of statements.[12]

This is a rather narrow "logical" view of critical thinking. Ennis continually seems to be assuming that there is a concept (or conception which we might ultimately be able to capture and explicate. A large portion of his 1979 presidential address consists of a listing and an elaboration of proficiencies (skills?) and tendencies such as observing, inferring, generalizing, and so on.[13]

Whether he intends it or not, there is reason to characterize at least some of his work as aimed at giving a conceptual analysis of critical thinking in terms of skills and tendencies.

In my terms Ennis has fixed the concept and standards of critical thinking. Thus he can adapt to the world only by changing goals or the world. Ennis is, however, sensitive to the variety of situations in which critical thinking, as narrowly defined, does not seem appropriate. Thus he speaks of the necessity for taking into account the pragmatics of critical thinking or the necessity of considering other points of view, or assessing the total situation and changing one's position when necessary. The way in which he seems to handle the difficulties when one cannot make the world conform to one's view of critical thinking, thus seems primarily to be to cast the net wider. It is not a "rationalization" in the pejorative sense I used in the math skills example, but it does involve restructuring the situation using alternative hypothesis, more evidence, and so on until the difficulty is removed. However Ennis' mode of adaptation does *not* involve altering the concept or standards of critical thinking.

John McPeck, on the other hand, criticizes the "basic skills" approach to critical thinking as represented by Ennis.[14] Essentially McPeck argues that the concept of critical thinking cannot be fully understood independently of substantive areas of thought. This is because these substantive areas of thought, the disciplines, give the background for deciding on hard cases. McPeck holds that there are many different conceptions of critical thinking, corresponding to the different canons of justification to be found in the disciplines. These modes of justification serve as the touchstone of good reasons for beliefs in any given area. McPeck's major ploy is to remind us that we are constantly being called upon to judge when preferred reasons are "good enough" and that such judgments presuppose an answer to the question, "Good enough for what?" Thus McPeck's epistemological criteria for critical thinking reinforce the notion that monitoring and evaluation will always involve a comparison of the given reasons with a set of perhaps changeable standards—the standards of the disciplines to which we should appeal in monitoring and evaluating our thought.

In my terms, McPeck's mode of adaptation consists of altering the standards of critical thinking depending upon the appropriate discipline in which the thought is to be located. Deductive logic with its self-contained standards is not sufficient for all of the areas in which we find critical thought. Nor have we been able to establish strictly logical criteria for inductive or informal logic or the scientific method in general. Rather the substantive standards of the disciplines must be used. Difficulties are handled by selecting the right standards which in turn are elaborated by the disciplines.

Next, consider Jane Martin who has recently been arguing that a male cognitive perspective has for too long dominated philosophy of education.[15] In part her argument amounts to suggesting that an ethic of care and nurturance must also play a role in a more complete understanding of cognition. People are *not* just reasoning machines, but affective beings as well, and the psychological and social must be given their just due.

More recently at a conference on critical thinking at Buffalo, Martin suggested that for Rousseau as well as for many moderns, critical thinking, as currently conceived, reduces real problems of the world to one dimension—the logical. The narrowly rationalistic education of Rousseau's Emile illustrates the extent to which the logical dominates the received conception of critical, rational thought. Yet Rousseau also dimly realizes, according to Martin, that the complete moral entity consists of Emile *and* Sophie—a rational aspect combined with a nurturing, caring aspect.

In my terms Martin's mode of adaptation suggests that not even recognizing the differing standards of the disciplines will always be sufficient to salvage critical thinking. We may, in some cases, have to consider expanding the concept beyond the logical, beyond the disciplinary, to the social, where care and nurturance and a sense of wholeness enter in. Difficulties may have to be handled by changing the concept.

An epistemology of adaptation makes all of the foregoing intelligible. We monitor and evaluate our thought and activity by noting the distance between the current situation and our concepts, goals and standards. We think critically when we tend to reduce that distance; but the distance can, in principle, be reduced by altering the situation, by altering the standards and concepts in terms of which the situation is assessed, or by altering the goals.

Ennis, McPeck, and Martin have each emphasized a piece of critical thought when viewed from the perspective of an epistemology of adaptation. Ennis reminds us that the concept of critical thinking is a central one, closely linked with logic and hard to change. McPeck points out the necessity of taking account of the substantive standards of the disciplines in assessing statements. Martin has challenged us to consider whether or not the very concept of critical thought might not need to be expanded. An epistemology of adaptation suggests that all might be ways of reducing the distance between situation and concepts, standards, and goals. Different tactics may be appropriate in different situations. True critical thought must be sensitive to all of these possibilities of adaptation.

III. Testing

In a recent article in the *Journal of Research in Science Teaching*, Morgenstern and Renner reported an analysis of a large sample of high school tests in the various sciences.[16] They compared the items on these tests with a list of ten "rational powers" found in the 1961 Educational Policy Commission's definitions of a hierarchy of cognitive skills. The list includes recalling, imagining, classifying, generalizing, comparing, inferring, deducing, analyzing, evaluating, and synthesizing. They found that the overwhelming majority of items required recall only. Indeed seven tests contained *only* recall items. It would seem that if we wish to test for the higher order cognitive skills, current tests will not serve us very well.

However, I want to argue that the technical requirements of standard "objective" tests, such as true-false and multiple-choice, render them almost unusable as tests for critical thinking, at least to the extent that monitoring and evaluating thought and activity is central to critical thinking.[17] The reason is simple. In a very real sense multiple-choice and other "objective" tests require an unambiguous right answer. Otherwise, the item is not acceptable on test-theoretic grounds. However, in order to get a single unambiguous answer, test constructors implicitly have to hold constant all but one of the various ways in which adaptive thought can occur. The test-item then will allow for only one way of reducing the distance between situation and concept or standard. Typically this can be done with some show of plausibility in only a very limited number of ways—one, limit the test to recall items; two, make the "right" answer a matter of well-accepted "deductive" techniques; three, give so much information that simple comprehension rules out the unwanted alternatives; or, four, put "trick" qualifiers on the wrong answers.

Furthermore, objective tests are at best only indicators of some underlying knowledge. Yet, as we all know, getting good test grades often becomes an end in itself. This situation occurs because the test results, conceived as standards indicating the knowledge, tend to become fixed. Students cannot demonstrate their knowledge except by doing well on the "objective" tests, even though these paper and pencil tests are only substitutes for the real competences at which we are aiming.

Recently, several psychologists have begun to question some of these artificial, yet long-accepted characteristics of paper and pencil testing.[18] Robert Sternberg at Yale and Howard Gardner at Harvard are two of the leaders in this regard.

Sternberg's theory of intelligence emphasizes the actual consequences of behavior in the world and the importance of those consequences for assessing intelligent thought. Gardner's theory stresses the need to see the development of intelligence in actual settings. This ecological strand in both psychologists' thought is quite compatible with the epistemology of adaptation I have been stressing in this paper and it underscores the notion that truly intelligent behavior can only be assessed in context. For example, Wagner and Sternberg say

> Intelligence does not operate in a vacuum, but rather in a world that is constantly increasing in complexity. If our understanding of intelligence is to have any relevance for understanding the interface between the individual and this world, it will have to study the functioning of the individual in this world, rather than merely in a laboratory or on a standardized test.[19]

Consistent with an epistemology of adaptation, this emphasis on assessment in context reminds us that there are in principle a variety of ways of adapting to real problematic situations and it is wholly inappropriate to fix on only artificial substitutes for real performance.

Consider the following example described by Gardner.

> ...Kpelle tribespeople in Nigeria rarely sorted spontaneously on the basis of a superordinate category. If given the names of a number of fruits, animals, and tools, the Kpelle would be expected to sort by these categories, rather than, say, by the functions that certain elements can serve. But these investigators found that the Kpelle tended to sort by function. Such sorting is nonpreferred in our culture and closely cognate ones and is considered indicative of a relatively lower level of cognitive development. Indeed, in vocabulary and similar subtests of I.Q. tests, functional definitions (an apple is "something you eat") are generally scored with fewer points than are taxonomic definitions (an apple is "a kind of fruit"). But what is of greatest interest in the story of the Kpelle is that the tribespeople could sort taxonomically, but did so only when the examiners, in desperation, asked them to sort the way stupid people would. Clearly, the Kpelle and their examiners perceived intelligent behavior differently.[20]

Gardner goes on to give another example provided by Seymour Sarason who described

> the rather bizarre situation that confronted him when he went to work at his first job, as a psychologist in a school for the mentally retarded. He arrived just as the students executed a successful escape from the school's restricted grounds. When the escapees were caught, Sarason was left to do his job; namely, to give the students the Porteus Maze Test! Curiously, the very students who had plotted and executed the successful escape were generally unable to complete even the first problem on the test. One must ask, which was the better measure of intelligence, the problem of escape or the Porteus Maze Test?[21]

Testing for critical thinking must in principle allow for the student to let us know how he or she structures the world and how the student proposes to deal with that world as structured. Sarason's inmates structure real escape from the institution quite differently than they do the paper and pencil exercise the psychologist attempts to impose upon them. Similarly, the Kpelle's needs for something to eat and how to get it appear to outweigh their needs for nice distinctions among what you eat. Under such circumstances only stupid people, indeed, would waste their time sorting into kinds of things to eat.

What, then does it mean to insist that intelligent behavior, critical thinking, and the like, can only be assessed in actual ecological contexts? My suggestion is that essentially we must turn current testing assumptions upside down. We must not conceive of ourselves as sampling the *outputs* of people's cognitive structures but rather as examining the ways that people deal with the *inputs* of the situations in which they find themselves. We must try to determine how subjects structure and actually deal with their environments rather than with how the tester interprets the subjects' responses.

Recall the feature of monitoring and evaluating one's action as one of the keys to critical thought. Inevitably, this monitoring and evaluating will be done *by* the individual in terms of his or her concepts and with his or her goals, purposes, and standards in mind. The essential element in this process will be how that individual structures or perceives his or her environment and what is done to bring the perceived environment into consonance with the cognitive structure. Such activities *may* be carried on in the same terms as those set by the tester, and in routine situations we can probably assume that they are, but they *need not be*. Especially in cases which call for critical thinking we need to know how *the student* monitors and evaluates, not how the tester wishes the student would monitor and evaluate.

The necessity of providing ways to find out how the subjects structure and deal with their environments indicates why we intuitively feel that such assessment instruments as essays and doctoral orals are so much better at really determining how well people think than are multiple-choice exams. An essay provides a setting in which the writer can reveal a fairly structured and possibly idiosyncratic way of dealing with the world. An oral allows the examiner to follow up on questions to see how the subject would deal with this variation or that.

Of course, I am not saying that it is impossible to assess critical thinking with multiple-choice tests, only that it is very difficult because of the way in which such tests preclude many of the possible modes of adaptation. One improvement we might try on multiple-choice exams would be to allow students, if they did not wish to answer a given question, to pick one of the choices given and write a stem that would make their answer true. Partial credit could be given for such an answer. Similarly, I am not claiming that oral or essay exams give full play to the variety of modes of adaptation, only that within the limits they set, it is sometimes easier to determine how well the student is dealing with the situation in the student's terms than by using multiple-choice tests.

The standardized "objective" test is primarily called objective because it can be easily and reliably scored by the tester.[22]

What could be a more obvious indication of a bias in favor of the *tester* and the *tester's* way of looking at the world? Of course, in some cases, we may have good reason to believe that the tester's way really is the only, the best, or at least the socially most appropriate way of looking at and dealing with the world. Factual recall, deductive logic and mathematics, and simple reading comprehension are examples. Thus for many school subjects, especially if critical thought is not emphasized, the important distinction between the tester's view of the world and the subject's view of the world makes no difference.

It is only as we begin to emphasize critical thinking and the higher order cognitive skills that the distinction between the tester's and subject's views of the world becomes important. It is in such cases that it becomes more apparent that there are a variety of ways of structuring experience and a variety of ways of closing perceived gaps between situation and concepts, goals, and standards. In such cases we are faced with a choice. On the one hand, we can defer to current conceptions of testing and insist that alternative modes of adaptation to their worlds by the subjects do not really exemplify critical thought at all. If we take this tack, we may at best find ourselves attempting to honor some marvelously adaptive behavior under a rubric other than critical thinking. At worst, we will label some very intelligent people as somehow stupid. Alternatively, we can insist that current testing dogma is quite limited and valid only in certain ranges. A notion of adaptation, properly elucidated, would lead to the development of new modes of assessment which can capture a more adequate idea of critical thinking.

An analogy is helpful here. Linda Darling-Hammond suggests the following one:

> Imagine for a moment that our nation's concerns about the quality and costs of health care resulted in the adoption of a single performance measure for judging patient health and doctors' competence. The cheapest, easiest, and most reliable measure of patient health we find is a simple, widely available tool called the thermometer. We decide to base all our health care decisions and rewards for doctors on patients' thermometer scores. After all, we reason, a patient with a good thermometer score is likely to be healthy in other respects as well.
>
> Three things happen in this scenario if our bureaucratic controls are effective. First, more aspirin is prescribed and consumed. Second, use of other treatment tools and methods declines because they are more costly and fail to show a direct, immediate relationship to thermometer scores. Third, doctors who are uncomfortable with the measure and its treatment implications become dissatisfied, complain about their lack of professional autonomy, and either engage in subversive practices or leave the established profession. Their complaints are dismissed as defensive and self-serving attempts to avoid accountability.
>
> Does health care improve? There is no way to know, because we have no other legitimate measures.[23]

Suppose that the thermometer had come to be the sole, or at least, main method of assessing health, much as the standardized test is the sole or, at least, main means of assessing cognitive competence. Faced with people who die despite good "thermometer scores" we could continue to insist that health really is just that which is assessed by our thermometers. We might want to talk about another feature of human existence which was somehow connected with dying when our temperatures were normal, but that feature would not be captured easily or at all in our health maintenance institutions, procedures or funding. Alternatively, we could recognize that thermometers may be only *one* assessment technique of an underlying notion of human health, and that we need to develop other measures if we are to have a more adequate conception of human health and how to promote it. Similarly, objective tests at best get at only part of what we mean by critical thinking. We must develop other modes of assessment if we are to promote a more adequate conception of critical thinking.

Adaptation can occur in a variety of ways and depends fundamentally on the individual's ability to find a congruence between the situation as structured by the student and the student's concepts, goals and standards. Thus, critical thinking must be assessed by means of understanding the ways the student views the world and how well the student can deal with the student's perceived reality rather than with the tester's view of the world. Of course, the student's and tester's views may coincide and we may in some cases as a matter of social policy wish to insist upon such a congruence. What we must remember, however, in the absence of absolutist foundations of knowledge, is that, *in principle,* true critical thought can always be directed at our basic principles themselves.

IV. Educational policy

There is no question but that testing drives educational policy to a very large extent. One might think, for example, of the "wall-charts" made popular by former Secretary of Education Terrell Bell, comparing states on the basis of their SAT and ACT scores. The Regents' Examinations in the State of New York not only provide a measure of individual performance, they are also used to single out schools needing special help. More and more states have instituted teacher examinations as part of their efforts to improve education through guaranteeing minimal teacher competence. Some states even rate schools and colleges of education on how well their students perform on these teacher exams.

Nor do we lack stories of the distortions introduced into the educational process because of the importance of tests. We all know about teaching to the test, but we have gone far beyond the Barron's-style books which collect old tests for study by students. There are stories of states where local school districts somehow manage to have field trips for large numbers of disadvantaged students on state testing days. I can only testify that in most schools in New York, some of the very best teachers will be teaching third and sixth grades. Those just happen to be the grades in which the state-wide Pupil Evaluation Program tests are given. Superintendents make no bones about raising scores in their districts through instituting specialized coaching programs. The list could be expanded indefinitely.

It is unlikely that the link between testing and policy implementation will be easily broken. Testing is a relatively cheap method for promoting certain kinds of educational policy. Although there are many educators who are convinced that current testing is wholly inappropriate for assessing higher order cognitive skills such as critical thinking, these same people will point to the enormous expense of instruments such as essays and oral examinations, two forms of assessment I noted as more in keeping with what would be required to test for critical thinking, properly conceived. Multiple-choice tests are economically feasible, they will say.

The short response to such cynics is, of course, to point out that thermometers are cheap, too, but they are poor tools for evaluating broken bones. X-rays work much better. Similarly, essays work better than multiple-choice exams for assessing critical thinking.

Another response is to emphasize that the professional judgments of teachers, working with children day in and day out on a variety of tasks and activities, are always available as a type of assessment. Teacher judgment is a test for which we have already paid. Indeed, as Wittgenstein earlier and Jim MacMillan last year have pointed out, agreements in judgments are the foundation of epistemology.[24] We could do worse than acknowledge the plain fact that *teachers* will inevitably make dozens of critical educational decisions every day. There is no way that we can avoid having teachers form their own judgments about the abilities of the children with whom they work. What we need to do is to find ways of improving those judgments and making them more reliable.

Teachers as testers do, of course, have their own interpretations of behavior. They are "subjective," it is claimed. That fact is often cited as a reason against using the kinds of open-ended tests I have favored for assessing critical thinking. However, the solution of introducing "objective" forms of testing misses the point. Objective tests ensure that, for the most part, all the *testers* will agree on the interpretation, but that does not answer the question of whether or not that interpretation is the one the *student* holds. Furthermore, it is the extent to which the *student's* concepts, goals, standards, and perceptions of the situation are adaptive that determines whether the student can be judged to be capable of critical thinking. Just as it is often said that hospitals seem to be organized for the benefit of doctors, so too do tests seem to be structured for the benefit of the tester.

The point to keep in mind is that the student's interpretation of a given situation *can* be determined empirically. We can hypothesize what the student's views are, we can introduce situations which, if we are right in our hypotheses, should lead to certain kinds of behavior, and then see if that behavior occurs. The method is not foolproof, but it can work. "Why do you believe such and such?" "What would happen if thus and so were to occur?" "How would you defend that claim?" All of the common sense questions and modes of assessment aimed at fostering what we ordinarily call understanding are, in essence, ways of finding out how other people do monitor and evaluate their situations. Teachers can be trained to utilize these methods much better than they have been to date. In short, there *are* alternatives to multiple-choice exams, if we would but make use of them.

What will happen to educational policy if we fail to revise our approaches to testing? At one level, things will go along pretty much as they have. Students who memorize easily and have good mathematical aptitudes will do well and will continue to have access to the better schools, colleges, and jobs. Many of these students can, and sometimes do, also pick up critical thinking almost in spite of themselves. So, we will probably muddle along.

We will not, however, make much progress in educating the large number of students who will need the higher order skills demanded by the twenty-first century. Even worse, we will surely condemn our society to a two-tier system—those who can pass multiple-choice tests and those who cannot. We will almost certainly lose the "street-smart" kids who do not knuckle under or do not do well on tasks involving recall and mathematical manipulations. Test scores instead of learning will become even more entrenched as the goal of schooling.

We will have denied ourselves the insights of broadened conceptions of intelligence such as Sternberg's and Gardner's ecological models. We will have limited critical thought to the narrowest kinds of mathematical and deductive reasoning. For those of us who want more from education, we will have given up one of our most potent justifications for curricular goals that go beyond the kind for which we can easily test with current methodology. We will have no defense for the accusation that we are "soft" when in reality our opponents are blind to the real world.

It is a dreary prospect, one that need not occur, if we can begin to articulate the case. And it is a case to be made by philosophers of education. Few other scholars possess the knowledge of philosophy, testing, and policy analysis in the very practical realm of education. We can and must sound the alarm. We must analyze the notion of critical thinking. We must understand and elaborate the metacognitive processes of monitoring, evaluating, and correcting our thought. We need to explore the extent to which our cognitive structures interact with the world to give us the experiences we have and the processes we have evolved for adapting to the world. We need to open our consideration of knowledge processes to a whole variety of influences not historically viewed as cognitive so that we can include the social and the nurturant. We need to understand the grounding of logic in its adequacy for dealing with the world rather than attempting to make the world adequate to logic. We must understand current test theory, both its strengths and its limitations. We must work with psychometricians to develop and validate the kinds of tests we believe must supplement multiple choice tests. In particular we need to emphasize the necessity of understanding the cognitive structures *students* bring to the test situation and how well those structures work rather than the cognitive structures imposed on the situation by the tester. We must help policy makers to understand the limitations of current testing dogma and the possibilities inherent in treating teachers as real professionals capable of rendering informed and complex assessments of their students' abilities.

We can address the problem of testing for critical thinking if we bring our critical thought to bear on some of the unquestioned assumptions of logic, testing, and educational policy. Some of those assumptions will need to be altered if we are to be able to deal with the complex world we face. True critical thought allows no less.

Notes

1. College Entrance Examination Board, *Academic Preparation for College: What Students Need to Know and be Able to Do* (New York: The College Board, 1983).
2. Ibid., p. 7.
3. Ibid.
4. Ibid.
5. Ibid., p. 9.
6. Ibid., pp. 9-10.
7. Robert J. Sternberg, "How Can We Teach Intelligence?" *Educational Leadership*, Sept., 1984, pp. 39-40. See also Robert J. Sternberg, *Beyond IQ: A Triarchic Theory of Human Intelligence* (New York: Cambridge University Press, 1984).
8. Arthur L. Costa, "Mediating the Metacognitive," *Educational Leadership*, Nov., 1984, pp. 57-62.
9. Richard K. Wagner and Robert J. Sternberg, "Alternative Conceptions of Intelligence and Their Implications for Education," *Review of Educational Research* 54,2 (Summer, 1984): 179-223.
10. Ibid., p. 186.
11. Hugh G. Petrie, *The Dilemma of Enquiry and Learning* (Chicago: University of Chicago Press, 1981). My analysis in the book was implicitly concerned with showing how, in fact, we can be said to monitor and evaluate our activity, including our thinking activity. I suggested a control system model that ties the perception of the situation clearly to the concept or goal in terms of which it is perceived. The perceived distance between situations and goal is conceived as the motive force driving the action to bring the situation closer to the goal.[1]
12. Robert Ennis, "A Concept of Critical Thinking," *Harvard Educational Review* 32,1 (Winter 1962): 83-111.
13. Robert Ennis, "Presidential Address: A Conception of Rational Thinking," *Philosophy of Education 1979 (Normal*, IL: The Philosophy of Education Society;, 1980).
14. John E. McPeck, *Critical Thinking and Education* (New York: St. Martin's, 1981).
15. Jane R. Martin, "The Ideal of the Educated Person," *Philosophy of Education 1981* (Normal, IL: The Philosophy of Education Society, 1982).

---

[1] Revised and expanded (2011) Menlo Park, CA: Living Control Systems Publishing

16. Carol F. Morgenstern and John Renner, "Measuring Thinking with Standardized Science Tests," *Journal of Research in Science Teaching* 21,6 (September, 1984): 639-648.
17. McPeck, p. 143ff.
18. See, for example, Robert J. Sternberg, "Testing Intelligence Without IQ Tests," *Phi Delta Kappan,* June 1984, pp. 694-698; and Howard Gardner, *Frames of Mind: The Theory of Multiple Intelligences* (New York: Basic Books, 1983).
19. Wagner and Sternberg, p. 187.
20. Gardner, p. 695.
21. Ibid.
22. See, for example, Norman E. Gronlund, *Measurement and Evaluation in Teaching,* 3rd ed. (New York: MacMillan, 1976).
23. Linda Darling-Hammond, "Taking the Measure of Excellence," *American Educator,* Fall 1984, p. 46.
24. C.J.B. MacMillan, "Love and Logic in 1984," *Philosophy of Education 1984* (Normal, IL: The Philosophy of Education Society, 1984).

[1992]
Interdisciplinary Education: Are We Faced With Insurmountable Opportunities?

Over 15 years ago, I wrote a paper titled "Do You See What I See? The Epistemology of Interdisciplinary Inquiry" (1976). In that article I discussed the relationship of interdisciplinarity to the disciplines and some of the features of the burgeoning field of interdisciplinary inquiry and education. Of course, a concern with a narrow focus on the disciplines predates my 1976 article. In the first part of the century, Dewey (1916, 1933, 1938) implicitly attacked a narrow formulation of the disciplines as the basis for education in his elaborate theory of the role of experience in learning. Similarly, the National Association for Core Curriculum, in its *Core Teacher* series, has been plumping for interdisciplinary education for over 40 years.

In the last 15 years, however, interest in interdisciplinary matters has increased significantly. The interest has been especially noteworthy in higher education. For example, Jane Roland Martin (1982) has questioned what she calls the dogma of God-given subjects in order to try to build a more adequate curriculum that does not rely solely on the traditional disciplines and that finds a place for such paradigm examples of interdisciplinary education as African-American and women's studies. Ernest Boyer (1987) has analyzed the undergraduate experience, criticizing the overemphasis on the disciplinary major to the detriment of general education. Most recently, the Association of American Colleges has been active in trying to bring coherence to the undergraduate college curriculum and has just completed a 3-year project on the academic major (1991). One of the working groups participating in the project focused on interdisciplinary studies, and its results have been jointly published by the Association of American Colleges and the Society for Values in Higher Education (1990).

Reproduced with permission of publisher from:
Grant, G. (ed.) Review of Research in Education, Vol 18. Washington, DC: American Educational Research Association.

Turning to K-12 education, John Goodlad (1983) has decried the sameness of schools and the deadening formality of instruction that is closely tied to the disciplines, especially in secondary schools. More recently, Ted Sizer's Coalition of Essential Schools (1984, 1988; Sizer, 1984) has had a significant impact on the educational scene with its slogan that "less is more" and the push for teachers as generalist coaches and students as workers. From a slightly different direction, Lauren Resnick, in her American Educational Research Association (AERA) presidential address (1987), contrasted the many differences between in-school and out-of school learning to the disparagement of much of the abstract, decontextualized, disciplinary nature of in-school learning.

In addition to these, as well as many other educational developments that have sharpened our interest in interdisciplinarity, two volumes have recently appeared that provide excellent summaries and discussions of the growing literature in the field. In 1986 Daryl Chubin, Alan Porter, Frederick Rossini, and Terry Connolly edited *Interdisciplinary Analysis and Research: Theory and Practice of Problem-Focused Research and Development*. This book contains an excellent collection of the most important articles in the field. *Interdisciplinarity: History, Theory and Practice* (1990a), by Julie Thompson Klein, is a superb comprehensive study of the concept of interdisciplinarity. Klein explores the definition of interdisciplinarity, examines the relationship of interdisciplinary studies to disciplinary work, and surveys the state of the art of interdisciplinarity in areas such as research, education, and health care. The book also contains an extensive bibliography on interdisciplinarity, as does the volume edited by Chubin, Porter, Rossini, and Connolly. Any serious study of the present-day status of interdisciplinary education should start with these two outstanding books.

Because of this recent summary work in the area, I am going to approach this review with, perhaps, a bit more focus on specific areas and questions than is customary in a *Review of Research in Education* chapter. It seems to me that the somewhat unbridled enthusiasm within education for interdisciplinarity could profit from a serious analysis of the concept and origins of interdisciplinary thought and from a critical study of the implications for interdisciplinarity of some of the major theses of current cognitive science. In this way, I will suggest that an enthusiastic but naive view of interdisciplinary education could indeed lead us to "insurmountable opportunities."

There are three main conceptual strands I wish to trace throughout this discussion of interdisciplinary education. These are, first, Aristotle's (1941) distinction between theoretical and practical wisdom; second, the distinction between the locus of interdisciplinary thought residing primarily in the individual and primarily in groups or other social arrangements; and, third, the modern constructivist view of learning, foreshadowed by Dewey's (1938) theory of experience.

Let me say a little more about each of these strands of thought. The Aristotelian distinction between theoretical and practical wisdom is the distinction between answering the question "What is the case?" and the question "What ought one to do?" The former is concerned with understanding the world as it is, the latter with acting in the world. Theoretical wisdom pursues the truth; practical wisdom pursues the good. Indeed, one way of viewing the task of any modern professional school such as law or medicine or education is to ask how the understandings provided by the basic disciplines presumably undergirding that profession are connected to the practice of the profession. How does jurisprudence or legal theory improve the practice of law? What is the relationship between biology and chemistry on the one hand and health on the other hand? How can a knowledge of psychology or sociology or history inform education?

The standard notion that such "theoretical foundations" are simply "applied" to the practical question of what ought to be done has come under increasing attack (Kennedy, 1987). The problem is that the concept of some more or less recipe-like following of rules derived from theoretical understanding simply does not work in any reasonably complicated activity such as teaching and learning. Although still popular in some policy quarters, the idea that we can directly apply the knowledge of the disciplines to develop specifications of what teachers should routinely do to get students to learn has pretty much been abandoned in the face of our increasing understanding of the complexities of the work of teaching and learning (e.g., see Schön, 1983, 1987; L. Shulman, 1986, 1987b).

The second strand of thought I wish to pursue is the distinction between the knowledge of an individual and the knowledge of a group. I am here referring to a wide range of ideas. There is the relatively simple notion that in many instances, simply summing the individual knowledge of the members of a group results in more knowledge than is held by any single individual in the group. There is the more sophisticated notion that knowledge, theoretical or practical, depends essentially on a community of inquirers (e.g., Benson, 1989; Hamlyn, 1978). There is also the observation that we sometimes actually build knowledge into a social system of some sort or other. Resnick (1987) cites the example of modern navigation on U.S. Navy ships. Additional striking examples for those who participate in them are electronic mail networks and modern electronic library searching techniques. There simply seems to be more and a different kind of knowledge residing in the group or system than resides in any given individual.

The third area is that of a constructivist view of learning. Under this conception, knowledge is not just handed over from teacher to learner. Rather, the idea is that learners construct meanings that enable them to make sense of the situations in which they find themselves. This view has its roots in Kant's (1961) notion of the synthetic a priori, Dewey's (1938) concept of experience, Kuhn's (1970, 1974) theory of scientific revolutions, Toulmin's (1972, 1977) analysis of the development of conceptual understanding, my own examination of the dilemma of learning and understanding (1981), and more recent accounts of psychological development (e.g., see Anderson & Pearson, 1984; Brown, Collins, & Duguid, 1989; Gardner, 1983; Phillips & Soltis, 1985; Sternberg, 1985). The core idea is that a variety of ways of dealing with the world and its problems may be appropriate, especially within the sometimes limited ecology of an individual, and individuals construct their own ways of making sense of their environments in accordance with past history and the details of current situations, both physical and social. Teaching and learning then become ways of trying out alternative sense-making strategies in terms of their increasing adequacy.

The idea of interdisciplinary teaching and learning is a powerful and appealing one. We must, however, be careful to examine its limitations as well as its promises. The three strands of thought just sketched will help us in thinking about the limits of interdisciplinary education. What, for example, is the difference between interdisciplinarity as a means of promoting theoretical knowledge and its use in practical problem solving? How are the disciplines used in these two areas? Can interdisciplinary knowledge be located in the heads of individual teachers and students, or must we think harder about the use of social arrangements to pursue interdisciplinarity? What implications does the idea of the construction of knowledge have for interdisciplinary studies? Does it mean that "anything goes"? Or are there some kinds of limits to the constructions? These are the major questions to be addressed in what follows.

A lexicon

Perhaps the seminal work in scholarship in the field of interdisciplinarity is the report of the First International Conference on Interdisciplinarity sponsored by the Organization for Economic Cooperation and Development. It is titled *Interdisciplinarity: Problems of Teaching and Research in Universities* (1972). The basic terminological distinctions were drawn in this work, and they enjoy a reasonably broad acceptance in the literature; however, given the state of development of the field, these distinctions are by no means universally accepted, and a number of other categorizations are also evident (Klein, 1990a).

Disciplinarity

Interdisciplinarity cannot be understood apart from the concept of disciplinarity. Roughly, the idea of a discipline today connotes a number of things, including:

- A specialization of knowledge within some sort of overriding unity of cognitive endeavor, such as the natural sciences (Toulmin, 1972)
- The fact that the unity of a discipline seems to come from (Hirst, 1974) a common set of core metaphors and concepts defining the field of inquiry, a particular set of observational categories for structuring experience in the field, specialized methods for investigation, a specification of the means for determining the truth or justification for claims made within the field, and, perhaps most important of all, an idea of the purposes to be served in investigating the field (e.g., in physics, the desire to understand the nature of the physical world in which we find ourselves)
- An organized grouping of people who study the discipline, train other practitioners, and form the social mechanism for arbitrating among varying truth claims within the discipline; this often involves university departments and degrees, national societies and conferences, and peer-reviewed scholarly journals and publications (Kuhn, 1974)

Clearly, there can be a variety of things that are called disciplines and a variety of arguments about whether something is or is not a discipline. A good recent analysis (Becher, 1989, 1990) describes the cultures that develop in and around the disciplines and points to the ways in which the disciplines are coming under increased scrutiny from a number of perspectives.

Multidisciplinarity

Given the above characterization of a discipline, the notion of multidisciplinarity is simply the idea of a number of disciplines working together on a problem, an educational program, or a research study. The effect is additive rather than integrative. The project is usually short-lived, and there is seldom any long-term change in the ways in which the disciplinary participants in a multidisciplinary project view their own work. Someone, perhaps a project manager, needs to glue the disciplinary pieces together, but that is all that happens by way of integration. Traditional distribution requirements in high school or college curricula are typically of this nature. Any integration is simply assumed to take place in the heads of individual students rather than there being a carefully thought-out system of general education. Health care and special education teams often operate in this mode. Klein (1990a, pp. 59-60) cites the middle phase of the Philadelphia Social History Project, where social scientists and demographers were added to the project's original historians, as an example of multidisciplinary work. It is *group* work rather than *team* work.

Interdisciplinarity

Interdisciplinary research or education typically refers to those situations in which the integration of the work goes beyond the mere concatenation of disciplinary contributions. Some key elements of disciplinarians' use of their concepts and tools change. There is a level of integration. Interdisciplinary subjects in university curricula such as physical chemistry or social psychology, which by now have, perhaps, themselves become disciplines, are good examples. A newer one might be the field of immunopharmocology, which combines the work of bacteriology, chemistry, physiology, and immunology. Another instance of interdisciplinarity might be the emerging notion of a core curriculum that goes considerably beyond simple distribution requirements in undergraduate programs of general education (Gaff, 1989; Newell, 1986, 1988). In many ways, the integrative thrust of interdisciplinary thinking is often a central feature of efforts to reform general education rather than a frill.

Turning to the schools, there are a number of national efforts, to be discussed in more detail below, to turn the "layer cake" (first biology, then topped by chemistry, then topped by physics) approach to American science education on its side. These efforts would require an interdisciplinary approach to teaching science since, at any given time, a combination of biology, chemistry, and physics would be studied. The various interconnections among these traditional disciplines would then need to be emphasized and fundamental principles, including mathematics, could be taught and learned more efficiently and effectively. Truly interdisciplinary special education teams might approach the special education student as a person with multiple needs rather than as a case to which different perspectives—all duly laid out in their hierarchical order of medical expert, psychologist, teacher specialist, and classroom teacher—can be brought. Klein (1990a, pp. 60-63) describes the later intention of the Philadelphia project noted above as interdisciplinary in that the results of the multidisciplinary research were to provide the basis for a holistic framework for studying cities. The fact that no such integrative framework ultimately emerged illustrates both the integrative idea of interdisciplinarity and the disciplinary paradox to be discussed below.

Transdisciplinarity

The notion of transdisciplinarity exemplifies one of the historically important driving forces in the area of interdisciplinarity, namely, the idea of the desirability of the integration of knowledge into some meaningful whole (e.g., see Kockelmans, 1979; Toulmin, 1982). The best example, perhaps, of the drive to transdisciplinarity might be the early discussions of general systems theory (Bateson, 1979; Boulding, 1956) when it was being held forward as a grand synthesis of knowledge. Marxism, structuralism, and feminist theory are sometimes cited as examples of a trans-disciplinary approach (Klein, 1990b). Essentially, this kind of interdisciplinarity represents the impetus to integrate knowledge, and, hence, is often characterized by a denigration and repudiation of the disciplines and disciplinary work as essentially fragmented and incomplete.

If we now look at these rough and ready distinctions through the lenses of the three conceptual strands I noted above, some interesting results emerge. First, consider the theoretical-practical wisdom distinction. Strictly disciplinary activities tend primarily to be concerned with theoretical understanding, while multidisciplinary activities, and perhaps even some interdisciplinary projects, are more concerned with practical results. Transdisciplinary activities, to be sure, tend toward addressing questions of theoretical understanding, especially those of the unity of knowledge, but the distinction between theoretical concerns and practical questions in interdisciplinary work seems worth making.

"The disciplinary paradox"

Consider, for example, Klein's (1990a, chap. 7) analysis of what she calls the disciplinary paradox. The paradox is essentially that, on the one hand, the fragmentation of knowledge into the disciplines leads to the necessity for interdisciplinary approaches, yet, on the other hand, interdisciplinary approaches to knowledge can only receive an epistemic justification from the established disciplines. This may, indeed, be a paradox if one restricts one's attention to theoretical knowledge. In such a case, the only solution would seem to be to try to construct some transdisciplinary notion of knowledge that encompasses all of the disciplines and their specific methodologies and provides an overall epistemic justification for knowledge claims.

However, if one also remembers that interdisciplinary approaches are frequently interested in practical wisdom and the solution of practical problems, then it is not clear that in such cases there is a paradox at all. As one of the early pioneers of interdisciplinary thought, Rustum Roy, put it, there is the "inexorable logic that the real problems of society do not come in discipline shaped blocks" (1979, p. 163). Thus, if the disciplines are sometimes irrelevant to our practical concerns, it is not at all clear that they can provide the only justification for good interdisciplinary work. At least in those cases where the problem is a practical one, the success in solving the problem can be taken as a justification.

This has important implications for such movements in education as the increasing emphasis on teachers as researchers (e.g., Cochran-Smith & Lytle, 1990) and the importance of teachers' craft knowledge (e.g., J. Shulman & Colbert, 1987, 1988). Insofar as these endeavors are taken as interdisciplinary ways of addressing practical problems in the classroom, it is not clear that they will require a disciplinary justification above and beyond their success in the classroom.

Second, if we are interested primarily in interdisciplinarity as it applies to individuals, we typically begin with the disciplines and how they are learned by individuals. The subjects of secondary school and the disciplinary majors of the college curriculum are our reference points. We yearn after some, perhaps unattainable, vision of transdisciplinarity that we wish all children could acquire in the name of general education. However, the cooperative and collaborative problem solving of groups of people working in multidisciplinary or interdisciplinary teams may provide a more adequate characterization of what is possible by way of general education. It may even be that current movements toward a kind of multicultural diversity as a curriculum goal are more realistic than imposing a single consensus where none is available. Possibly, the only unity to be found in a plurality of cultures may be that they all, in their own ways, solve some of the very situation-specific practical problems with which each culture is faced. Furthermore, the coherence desired of general education would then become a coherence of seeing the diverse ways in which human beings can solve their practical problems of what to do.

Consider in this regard the implication of a constructivist theory of learning for interdisciplinarity. This implication is nicely drawn out by Jane Roland Martin (1982) in her discussion of two dogmas of curriculum (also Birnbaum, 1969). One of these dogmas is what she calls the dogma of God-given subjects—that the subjects and disciplines are, somehow, natural and God-given. It is Martin's contention, reinforced by the notion that meaning is constructed, that these subjects are chosen and constructed rather than handed over. They are chosen for some human purpose and can be more or less useful in serving that purpose.

There is, of course, probably an historical justification for most of the subjects and disciplines studied in school. These disciplines probably do codify reasonably widespread and useful human activities in that they have, at least in the past, given us the solution to a number of human problems. However, it is important to realize that they can be challenged and that they can be replaced with other subjects and disciplines. Thus, discussions of and arguments about women's studies, African-American studies, or Western civilization are to be expected and welcomed. If our disciplines and subjects are created rather than discovered and their usefulness lies in their ability to help us deal with the world in which we find ourselves, then curriculum will always be an essentially contested domain. Interdisciplinarity, in this sense, begins to look more and more like a way of trying to see and deal with the world in new and different ways that may be more adequate than the traditional ways.

The preceding discussion contrasting problem-solving approaches with epistemological issues of warranted belief echoes a related set of distinctions that runs through the literature on interdisciplinarity. This is the distinction between "bridge-building" between relatively firm, independent disciplines and "restructuring" by changing parts of the interacting disciplines (Group for Research and Innovation, 1975). It is the difference between "instrumental borrowing" to solve problems and "integration" or "synopsis" to achieve a new conceptual unity (Klein, 1985; Landau, Proshansky, & Ittelson, 1962; Taylor, 1969).

It is the disparity between a "vacant" interdisciplinarity that simply looks for commonalities among existing frameworks and a "critical" one that rethinks the nature of knowledge (Kroker, 1980; Robbins, 1987). Finally, there is the distinction, on the one hand, between traditional disciplinarity and its associated interdisciplinarity, which pretty much accepts the disciplines as given, and, on the other hand, the postdisciplinary and deconstructionist critique of disciplinarity, in which a radical interdisciplinarity or cross-disciplinarity becomes a new kind of theoretical imperative (Brantlinger, 1990; Elam, 1990; Fish, 1989). Of course, from the radical deconstructionist perspective a truly critical interdisciplinarity is, in the end, impossible because no approach to knowledge has any justification. Such an argument, however, seems plausible only if one ignores practical wisdom and the justification derived from successfully pursuing our practical problems.

In my 1976 article I argued that the minimal necessary conditions for successful interdisciplinary work were that the participants understand the observational categories and meanings of the key terms in each other's disciplines—pointing out that different disciplines often see the same things differently and mean different things by the same words served, then, to illustrate reasons why interdisciplinary work is sometimes so difficult. We seldom take account of these differences.

What has happened is that the deconstructionists (e.g., Fish, 1989) have been impressed with the fact that different people see the same thing differently and mean different things by the same words and have concluded that warranted knowledge is, in principle, impossible. What they seem to ignore, however, is the centrality of the practical in interdisciplinary research and education. We do solve problems, of both a research and an educational nature, more or less well. Thus, pointing out that a given theoretical perspective with its associated observational categories is only one of several ways of looking at the world does not imply that we cannot make judgments of better and worse in comparing those ways of looking at the world.

Many scholars appear to have accepted the tentative nature of the disciplines and the need for interdisciplinary activity. However, as the disciplinary paradox suggests, we have only begun to look at ways of warranting knowledge beyond those contained in the disciplines (e.g., Giroux & McLaren, 1986). Interdisciplinary work, with its emphasis on the practical, may be a resource for correcting this deficiency.

Both the problem-solving and unity-of-knowledge perspectives have historically driven movements toward interdisciplinarity (Geertz, 1980; Klein, 1985). However, the motivations and goals have been quite dissimilar (e.g., see Turner, 1990). The former is aimed at practical problem solving in terms of current ways of understanding problems, while the latter focuses on the warrants for knowledge (i.e., on transdisciplinarity). The deconstructionists emphasize the lack of any absolute grounding for traditional disciplinary knowledge claims and, still captured by the vision of transdisciplinarity, seem driven to conclude that knowledge is impossible. There is no absolutely privileged place to stand from which to evaluate knowledge claims.

But all of this discussion actually depends on our continuing to draw a sharp distinction between theoretical and practical knowledge and assigning the disciplines solely to the realm of theoretical knowledge. If, however, we think of thought and action as much more closely connected, it can be argued that the disciplines themselves really are, or should be considered primarily as, means of solving practical problems of what ought to be done. In this sense, the disciplinary paradox can be solved not only for those cases in which the interdisciplinary work is focused on practical concerns, but also for the cases in which the justification for integrated, interdisciplinary systems of knowledge is being sought.

Throughout the history of ideas, the traditional disciplines have generated any number of theoretical problems of understanding that by now often have little connection to the ordinary problems of human existence. What is required is not some absolutely privileged standpoint from which to articulate a warranted notion of transdisciplinarity, but a return to the roots of the disciplines themselves as organized means of dealing with the world in which we find ourselves (e.g., see Casey, 1986; Geertz, 1980, Simons, 1989, 1990). As such, the worth of disciplines or interdisciplines can be judged on roughly pragmatic grounds. Both the traditional disciplines as well as new interdisciplinary and transdisciplinary ways of thinking are to be judged on the basis of how well any of these constructed systems of thought enable us to solve the historically conditioned, yet constantly evolving, practical problems of living (Petrie, 1981).

Having provided a lexicon, discussed the disciplinary paradox, and argued that the two imperatives for interdisciplinary work—the practical and the theoretical—can be understood as related, I want, in turn, to investigate interdisciplinary research and the growth of knowledge, the idea of interdisciplinary problem solving, and, finally, interdisciplinary approaches to education, both in higher education and in the schools. I hope to make clear that these areas are not separate and that preconceptions in one area often affect what happens in another.

Interdisciplinary research and the growth of knowledge

Donald Campbell (1969) has provided us with a compelling metaphor for interdisciplinary research and the growth of knowledge. In his paper, "Ethnocentrism of Disciplines and the Fish-Scale Model of Omniscience," Campbell argues persuasively for a social view of knowledge. Indeed, one of his main contributions in the paper is to note that even within the disciplines themselves, the integration and comprehensiveness we find is a collective product and not the accomplishment of any one scholar. Every scholar within any discipline has slightly different experiences and slightly different expertise. This fact actually follows from the notion of the construction of meaning by individuals from their unique experiences.

How, then, do we achieve the notion of a single discipline? Campbell's (1969) argument is that it is through the overlap among disciplinarians that a collective communication, competence, and breadth is attained. We have enough common experience to allow for the more or less common observational categories, more or less common methodological approaches, and more or less common core metaphors involved in defining a discipline (Petrie, 1976, 1981). In turn, these commonalities are roughly accounted for by the fact that as human beings living in the same world, we have similar basic problems to solve in terms of coping with that world. Thus, some commonality within individual disciplines is to be expected.

However, if one conceives of different subsections of the world as more or less well known by different individuals, much as a fish is covered by overlapping fish scales, then the problem with the disciplines is that the disciplinarians within a single discipline cluster much closer together than do the scales on a fish. It is as if small portions of the fish were covered by a large number of overlapping fish scales, but between these clusters, significant portions of the fish are uncovered. Campbell (1969) remarks how the social inventions of university departments with journals and tenure and the whole set of academic rituals actually work toward keeping the individual clusters closer together and against any individual's covering the spaces between the disciplinary clusters.

Interestingly, Campbell (1969) uses his social model to argue that rather than trying to create interdisciplinarians who truly know both disciplines (they would have to be equivalent to very large fish scales), it is more plausible to encourage individuals to become quite knowledgeable about small parts of the world that are not yet well known but that can overlap, at least to some degree, with a discipline that is already well known. In this way interdisciplinary work can move forward. Indeed, as these new areas are explored, it may even be that the "center of gravity" of a discipline may shift and different parts of the world may come to be more densely covered by the individual researchers working in the new area. Thus, geography can be seen as a movement away from drawing maps of the world toward a study of how we use space.

Of course, devising social structures that will allow for this kind of communication is not easy, especially in educational institutions. In higher education (e.g., see Caldwell, 1983; Rich & Warren, 1980), the traditional departmental structure tends to keep the disciplinarians quite isolated, although professional schools and area studies break down the gaps to some degree. In the schools, the problems of isolated classrooms and their inhibiting effects on communication among the professionals are well known. Indeed, this feature lies behind much of the reform-oriented analysis of the Holmes Group (see *Tomorrow's Teachers*, 1986, and especially *Tomorrow's Schools*, 1990). Fortunately, much general institutional and organizational analysis today (Peters & Waterman, 1982, continues to be a key reference) focuses on the need to build organizations in which there is a good deal less specialization (read "disciplinarity") and a good deal more integration (read "interdisciplinarity").

Viewing the disciplines and the movement toward interdisciplinary work as the result of social processes has a certain liberating effect on thinking about the growth of knowledge. On the one hand, the idea of a social grouping with more or less similar ways of seeing the world accounts for the strength of the notion of a community of inquiry in thinking about the growth of knowledge. People can communicate with each other, even given their more or less idiosyncratic experiences, if they are all working in roughly the same area of knowledge. Such a relatively stable social grouping explains why disciplinary work is thought to be so important. At the same time, it shows that some problems do not get addressed if these problems lie outside the interests of the dominant disciplinary group.

Campbell (1969) uses this idea to explain why university departments are so powerful and why they must sometimes be resisted. The fish-scale model would also explain the development of area studies in the social sciences and the more recent emergence of African-American studies, women's studies, and the like (e.g., Miller, 1982). Individuals are trying to cover new parts of the intellectual landscape, yet to be understood at all, they must have some ties to the traditional disciplines. However, as their studies progress, they may be able to develop new groupings of scholars in these fields themselves, rather than relying on the traditional disciplines (Clifford & Marcus, 1986).

This is reminiscent of Klein's (1990a) disciplinary paradox and Martin's (1982) dogma of the God-given subjects. In each case one can escape the straitjacket of the disciplines by looking for new and emerging human problems that need to be solved outside of the disciplinary ways of looking at things. At the same time, it explains the inherent difficulty of interdisciplinary research and the innate conservatism of the disciplines. People do tend to experience the world in the ways in which they were taught, and it requires major efforts to break out of the mold.

The theme of theory and practice is relevant here as well. The logic of interdisciplinarity, as laid out by Campbell (1969), suggests that sterile theory without the anchor of problem solving will keep the disciplines fragmented and incomplete. Thus, as Klein (1990a) has also noted, interdisciplinary thought often emphasizes the reunification of the theoretical and the practical. This movement toward unification has been exemplified recently for those of us in education in powerful critiques of the nature of schools and colleges of education (Clifford & Guthrie, 1988; Judge, 1982). The crux of the issue is the extent to which schools and colleges of education at the university level can conceive of themselves as true professional schools, committed to the practical needs of the profession as opposed to succumbing to the pull of the traditional disciplines that are enshrined in the departments of the university.

As a 10-year veteran dean of a graduate school of education and one of the founding members of the Holmes Group, I can personally testify to the importance of this issue. Our ability as schools of education to come to grips with the practical question of our place in the university will determine whether or not we will survive. A metaphor for the change I am advocating is the nearly yearlong debate we had at the State University of New York at Buffalo about whether or not to change our name from the "Faculty of Educational Studies" to the "Graduate School of Education." In the end, the Graduate School of Education was accepted, but not without considerable discussion of the nature of scholarship and the relationship of the theoretical and the practical. In addition, the 1991 national conference of the Holmes Group was devoted to the question of the nature of inquiry in the new, more collaborative relationships being proposed between schools of education and educational practitioners (Holmes Group, 1990).

The constructivist theory of learning and knowledge development is also quite congenial to interdisciplinary ideas of knowledge growth and development. Knowledge structures (i.e., disciplines) are constructed in response to certain general human needs and problems. The structures are built with certain purposes in mind and can be judged as more or less adequate depending on their long-range ability to provide solutions for the characteristic problems they were built to address. But as new problems arise (e.g., the growing awareness by minorities and women that their place in society has been largely defined for them), new, interdisciplinary areas of study may arise (Kolodny, 1984; Stimpson, 1988; Stimpson & Cobb, 1986). An excellent article by Brian Turner (1990) analyzes the sometimes conflicting reasons for pursuing interdisciplinary research and education in the medical field. The question is whether or not to accept the world as described currently and undertake interdisciplinary work to solve problems as currently defined or to use interdisciplinary investigation to challenge the fundamental conception of the traditional ways of defining our disciplines and problems. In such cases, however, there is, as Klein (1990a) puts it, always a "burden of comprehension" placed on the interdisciplinary advocate. The interdisciplinarian must continue to be able to communicate with the disciplinarians while standards of rigor and trustworthiness are being developed in the new area.

Ethnography is a good recent example in education of a field of study that had to face the criticisms of the disciplinarians as it struggled to find its own methods (Eisner, 1981; Firestone, 1987; Guba & Lincoln, 1988; Howe, 1988; Howe & Eisenhart, 1990; Jacob, 1988; Peshkin, 1988; L. Shulman, 1988). It was attacked in education as being both eclectic and lacking in rigor (Phillips, 1983, 1987a, 1987b), but, of course, those charges are themselves dependent on a particular, generally positivist view of science and disciplined inquiry, a view that is precisely under question. The danger is to fall back on the accepted disciplinary standards of evaluation rather than to develop the new standards appropriate to the new questions being asked. However, keeping the constructivist view clearly in mind helps one to avoid the temptation to dismiss new interdisciplinary proposals out of hand. All of knowledge is constructed, including the current disciplines, and the ultimate test of their worthwhileness is whether or not they allow us to deal more adequately with the world, physical and social, in which we find ourselves (Petrie, 1981).

Interdisciplinary problem solving

The area of interdisciplinary problem solving has already been referred to in Roy's (1979) famous quotation that societal problems do not, at least nowadays, come in discipline-shaped blocks. This idea is exemplified in any number of other examples—from the interdisciplinary development of the atomic bomb during World War II to the most recent war on drugs and the debate over the adequacy of our nation's concern for children. In each instance, the problem that has been identified seems to call for an interdisciplinary solution. In most cases the solution tends to be what I have called "multidisciplinary." That is, a mission-oriented team of experts from a variety of disciplines is assembled and given the charge of solving the problem. A substantial literature has grown up around the problems of setting up, organizing, and running interdisciplinary problem-solving groups (Chubin et al., 1986; Klein, 1990a, chap. 8).

As might be expected, there are a number of difficulties with such undertakings. Key among these have to do with different modes of structuring their experience by different members of the team and the blind spots this causes. After all, if a given discipline has as part of its essential makeup a certain way of seeing things (Petrie, 1976), then that way of seeing things will automatically also be a way of *not* seeing things. This perceptual feature of interdisciplinary inquiry still seems to me to characterize one of the necessary conditions for successful interdisciplinary work. One must be able to see the world in the ways in which the other members of the interdisciplinary team do. It takes a certain kind of broad-gauge scholar (e.g., one who is not worried about obtaining tenure or who already has it) to be able to experiment with new ways of looking at and conceiving of things. The problem also needs to be clearly stated in relatively nontechnical language, and the group needs to have leadership sensitive to the many different ways of thinking about the issue.

The construal and definition of the problem to be solved is one of the key features of interdisciplinary problem solving (Chubin, Porter, & Rossini, 1986). How should our experience be structured so as to permit the greatest chance of coming up with a solution? Should we view the drug problem as one of interdicting supply? As one of decreasing demand? As one of social anomie? As one of building a character sufficiently strong so that one can "just say no"? All problem solving in Dewey's sense (1933) is interdisciplinary. How can we structure our experience so that inquiry can lead to solutions of our problems along with continued growth and education? The justification of both the traditional disciplines and any new interdisciplinary study must be that they ultimately allow us to pursue our purposes in an ever-changing but broadly stable world (Petrie, 1981).

Porter and Rossini (1984) have offered the very useful STRAP framework for thinking about interdisciplinary problem solving. The framework consists of analyzing the Substantive knowledge required to solve the problem, the Techniques needed, the Range of substantive knowledge and techniques to be used, the Administrative or organizational complexity required, and the Personnel needed. This approach offers a more systematic updating of the features I mentioned in 1976. The framework also throws into clear relief the salience of the theoretical-practical distinction, the question of individual or group expertise, and the construction of knowledge issues that I am pursuing.

The theoretical disciplines are pressed into service only insofar as they appear to cast some light on the problem to be addressed. In this sense, interdisciplinary problem solving is a paradigm case of exploring, on the one hand, the relationships between theoretical understandings of why something is the case derived from codifications of substantive knowledge, and, on the other hand, technique and practical questions of what ought to be done. The justification for a given discipline in interdisciplinary problem solving is how well it assists in solving the problem. Notice, too, that the notion of the "application" of theory to practice in interdisciplinary settings has a hollow ring to it. Only if a given problem falls wholly within a discipline does the notion of applying theory to practice even appear to make sense. The more typical case in mission-oriented interdisciplinary research is that the various disciplines offer different scenarios and point to different, possibly salient features of the problem situation, but the solution is not a disciplinary matter.

For example, the recent calls for establishing the school as a center for integrated social services for at-risk children (e.g., Kirst, 1991) can be conceived of in at least two ways. First, it might be that the school simply happens to be where the child is for a good portion of the day, so that on logistical grounds alone, that is where he or she should be treated medically, counseled psychologically, and remediated socially and educationally. An interdisciplinary approach to the problem, however, might look at the situation not from the point of view of the various disciplinary experts, but rather from the point of view of the overall welfare of the child. In the latter case, while the doctor, nurse, psychologist, social worker, and remedial teacher may all have something to say about the situation, none of their individual "solutions" will be primary. Nor is this simply a bland sort of eclecticism. It is, instead, a structuring of the problem in a way that gives hope of a real solution.

Interdisciplinary problem solving also illustrates very clearly the difference between conceiving of interdisciplinarity as located in the individual and conceiving of it as located in the group. Almost all interdisciplinary problem solving occurs in groups, groups that must be organized and administered. With respect to personnel, we still have the occasional individual who can master several disciplines, but as Campbell (1969) pointed out, such persons are rare indeed. It seems much more promising to look at the problem as one of blending the correct social and organizational means of interaction and integration. Collaboration, rather than competition, must be the watchword.

The implications of interdisciplinary problem solving for education are several. First, insofar as the calls for curricular revision (discussed more fully below) emphasize problem solving and "higher order thinking skills," as many of them do, the STRAP framework provides a good way of thinking about these suggestions. What is the appropriate substantive knowledge? The techniques? How much is required? How should we organize the delivery of the new curriculum? What personnel should be involved? Single teachers? Groups of teachers? Others?

Second, the literature on teacher thinking and problem solving (e.g., Berliner, 1986, 1989; Kennedy, 1987; Schön, 1983, 1987; L. Shulman, 1986, 1987a, 1987b) can benefit from the work on interdisciplinary problem solving. After all, the work of the typical teacher, especially the elementary school teacher, is fundamentally interdisciplinary. One of the more interesting implications is that team teaching, in both elementary and secondary schools, may be a most useful way to conceive of approaching complex curricula. Instead of the nearly impossible task of trying to devise elementary teacher education programs to put all of the necessary knowledge in the head of a single teacher (Petrie, 1987), one can perhaps organize elementary teaching as if it were an interdisciplinary problem-solving group.

Interdisciplinary education in colleges and universities

Interdisciplinary education takes a number of different forms in higher education (Klein, 1990a; Levin & Lind, 1985; Organization for Economic Cooperation and Development, 1972). According to Klein (1990a, chap. 10), interdisciplinary education is found primarily in a few revolutionary institutions in which the entire program is conceived of as interdisciplinary, within so-called "area" studies at traditional universities, in liberal or general education programs, as a minor within a disciplinary structure, and within at least some professional programs.

Klein (1990a, pp. 157-163) mentions several institutions, primarily abroad, that have attempted to base the whole of their educational programming on interdisciplinary principles. All attempted to find a transdisciplinary definition of knowledge and knowledge acquisition. It is important to note that none of the institutions remains the same today as when it was founded. All have had to accommodate to a version of the disciplinary paradox in which the prevailing disciplinary mold puts enormous pressure on them to conform, from recruiting faculty to finding jobs for their students (Trow, 1984/1985). All, however, retain some important vision of interdisciplinarity within their mission.

Area studies, on the other hand, have a seemingly firm foothold in the modern university. Early geographical area studies, such as African, East Asian, and the like, blazed the way for more recent additions such as African-American studies, women's studies, American studies, and programs in science, technology, and society. Sometimes these programs borrow faculty from traditional disciplines; in other instances they have their own departments and degrees. Very often they are heavily associated with the institution's general education program. Professors who work in such programs frequently have to deal with a particularly virulent form of the disciplinary paradox in that they are usually held suspect by their disciplinary colleagues, yet must not succumb to the temptation simply to fall back into disciplinary ruts and modes of thinking.

Liberal or general education provides another important site for interdisciplinary education in the colleges and universities (Boyer, 1981, 1987; Clark & Wawrytko, 1990; Gaff, 1989; Newell, 1986, 1988). Typically, the impetus toward general education is not initially problem based, but rather arises from the drive toward integrative knowledge. The traditional distribution program is at best multidisciplinary, with any integration simply assumed to take place within the heads of individual students. Core curricula, on the other hand, usually involve at least some integration in the heads of the faculty who plan them and very often also include specially designed core courses that sometimes cross disciplines and nearly always aim at some sort of integrating function (Undergraduate College, 1990). Often there are also special projects and experiences and "capstone" seminars to pull it all together. The key feature of a truly interdisciplinary general education program is, ultimately, the extent to which the program itself attempts to synthesize the elements of the curriculum instead of simply leaving it to the students.

A particularly useful interdisciplinary model for implementing general education is provided by Hursh, Haas, and Moore (1983). They use the developmental theories of Dewey (1916, 1933, 1938), Perry (1968, 1981), and Piaget and Inhelder (1969) to argue for a "skills-based" general education curriculum—one that focuses on problem solving. They stress Perry's notion of acting on beliefs, another theme of interdisciplinarity. The model also emphasizes multiple perspectives on the problem and the salient concepts and the necessity for analyzing the strengths and limitations of various disciplinary approaches.

Hursh et al.'s answer (1983) to those who hold that an emphasis on skills alone ignores the critical role of substantive knowledge in attaining intellectual skills (e.g., McPeck, 1981) constitutes a powerful argument in favor of interdisciplinary skills. Essentially, they grant that cognitive development depends on the acquisition of substantive knowledge but point out that this begs the question of *what* substantive knowledge. Insofar as substantive knowledge depends on ever more specialized investigation, we become wholly dependent on the (disciplinary) experts in these specializations. What we need, however, is an ability to critically evaluate the claims of the "experts." In short, we need general intellectual skills, obtained through interdisciplinary education, to know when to trust the experts. Although they do not address the issue, I believe that Hursh, Haas, and Moore still feel that this capacity can be generated within each individual student. If, however, the locus of interdisciplinary wisdom might be better conceived of in the group, there still would be reason to follow a skills-oriented, problem-solving general education program, only with significant emphasis on learning group problem-solving skills as well.

Interdisciplinary minors are another feature of the modern university. They represent a relatively innocuous concession to interdisciplinarity on the part of the strong disciplinarians who dominate most universities. Such minors are frequently problem focused, with an attempt to show how the disciplinary majors can be used in these areas.

Another important site for interdisciplinary education in the university is the modern professional school (Schön, 1983). Many graduate and undergraduate programs in the professions of medicine, law, social work, engineering, and education attempt to provide integrated experiences for their students in order to help them situate themselves within the larger problematic of their professions. My earliest experience with interdisciplinarity was in a school of engineering with a program devoted to humanizing the education of engineers (Petrie, 1976). I have also mentioned the debate within education (Clifford & Guthrie, 1988; Judge, 1982) regarding the extent to which graduate schools of education need to broaden their programs beyond simply conceiving of themselves as populated by applied disciplinarians, without at the same time succumbing to the temptation to become modern versions of the old normal schools (i.e., without succumbing to the disciplinary paradox).

My own background over the past 10 years as dean of a graduate school of education in a major public research university is instructive. Klein's (1990a, p. 131) description of the qualities needed to lead interdisciplinary efforts struck a responsive chord with me. The need for previous interdisciplinary experience, sensitivity toward different paradigms, commitment to problem solving, group interaction skills, and enormous energy and patience conforms extremely well to my experience. Among the most difficult tasks a dean of education has is bringing together the various educational disciplinarians around common problems in ways that allow them to continue to value their cultural and disciplinary backgrounds. The interesting thing is that the "disciplinarians" needed are not only the educational psychologists, sociologists, philosophers, historians, and curricular specialists, but also the academics from the rest of the university and the very practically oriented professionals from the field. As the literature predicts, our most successful efforts have occurred when there is a clear problem, or problems, to be solved and when the participants take the time to understand each other's backgrounds and ways of looking at and dealing with the world (Chubin et al., 1986; Petrie, 1976).

The distinction between theoretical knowledge and practical knowledge finds expression in a number of places in college- and university-based programs of interdisciplinary education. The revolutionary institutions and at least some programs of general education tend toward an integration of theoretical knowledge, although there is often a strong, if not well-articulated, sense of the primacy of the practical in current discussions of liberal education. Area studies, interdisciplinary minors, and professional programs, on the other hand, tend to recognize the essentially practical focus of their activities and, thus, seem to be more problem oriented. The truly interesting questions cluster around the relationships between the traditional core of the academy, with its emphasis on theoretical understanding, and the increasing social and professional pressures to be relevant to society's problems. One critical question for future interdisciplinary research might well be the relationship between theoretical understanding and practical wisdom.

Ernest Boyer (1990) has recently begun to address this problem with his attempt to articulate a broadened definition of scholarship and its different relationships to knowledge. He and his colleagues have begun to speak of the discovery of knowledge, the synthesis and integration of knowledge, the application of knowledge, and the presentation and representation (teaching) of knowledge. Scholarship and research might well focus on any one of these relationships. We in the Graduate School of Education at Buffalo have begun to address a broadened notion of scholarship in promotion and tenure guidelines. Using the concept of "professional service" articulated by Elman and Smock (1985), we are attempting to fashion a category of equal prestige with traditional concepts of research and teaching. The category of professional service is a blending of the applied research and professional interpretation of theoretical scholarship for the field with the kind of service on a national or state commission or study group that may result in new and influential standards for the field. Such attempts within the university to articulate the relationships between theoretical and practical knowledge will only increase in the future.

The question of whether the locus of interdisciplinarity is to be found in the individual or the group also finds significant expression in higher education. Current attempts to redefine a core curriculum appear to assume that we really can create interdisciplinary individuals, if only we get the ideas and integrating concepts right (e.g., see Clark & Wawrytko, 1990; "General Education," 1989; Interdisciplinary Studies, 1978). On the other hand, the critique offered by multiculturalists raises some significant questions regarding our ability to define a core curriculum. Implicit in the multiculturalist argument is the claim that there are a number of different situation-specific ways of dealing with the problems of life and that none is necessarily better than any other (Kolodny, 1984; Stimpson, 1988). The multicultural curriculum would acquaint students with that fact and some cultural examples and then celebrate the diversity.

Others find this approach to be hopelessly relativistic and incoherent (e.g., D'Souza, 1991; Hirsch, 1987).

It is not at all surprising, given the constructivist approach to learning, that there now rages a significant debate regarding the nature of general education. There are those who believe that traditional subjects are simply "given" and any retreat from them would represent a retreat from standards (D'Souza, 1991; Hirsch, 1987). Even some who accept the constructivist point of view argue for the superiority of classical conceptions of knowledge (Ravitch & Finn, 1987). Others point out that great chunks of the experience of certain people have not only been omitted from the classics, but the classics have been used to oppress those people (Stimpson & Cobb, 1986). The constructivist point of view clearly indicates that such debates are to be expected and even welcomed as we try to achieve increasingly better ways of dealing with the world in which we live. It is not, however, the case that accepting a constructivist position on knowledge acquisition necessarily leads to a kind of relativism. As noted above, notions of better and worse ways of dealing with the world still seem to have application.

Interdisciplinary education in the schools

Educational programs in the schools, especially the secondary schools, often take their substance from programs in the colleges and universities. The influence of the traditional disciplines tends to extend with an iron hand into the secondary school curriculum, so it is not surprising to find similar debates within the schools. One good example is the recent furor caused by the publication in New York State of a report titled *A Curriculum of Inclusion* (Commissioner's Task Force on Minorities: Equity and Excellence, 1989). This report to the state education department argued that the state social studies curriculum had omitted and denigrated the contributions of many of the minority cultures represented in the state. As might be expected, people lined up on both sides of the issue. Like similar debates over the canon in higher education, this debate raises a number of issues of interdisciplinarity regarding theoretical versus practical knowledge, the construction of meaning, and the disciplinary paradox. The original report was followed by another, *One Nation, Many Peoples: A Declaration of Cultural Independence* (Social Studies Syllabus Review and Development Committee, 1991), produced by a distinguished panel of scholars, teachers, and laypersons. This second report, although still quite controversial, appears to be much more in line with the view I have been arguing that there are many ways of solving the practical problems with which human beings are faced and it is important for students to recognize that fact and approach all cultures with a sympathetic but critical eye.

Aside from the influence of the university, however, there are a number of developments in K-12 education that raise significant issues of interdisciplinarity and that deserve to be addressed independently. In order to do this, I have chosen three areas to discuss in some detail—the recent call for new standards in mathematics and science education, Ted Sizer's Coalition of Essential Schools, and Lauren Resnick's (1987) critique of school learning contained in her AERA presidential address.

New standards for mathematics and science education

Let me turn first to the proposed reforms in mathematics and science education (American Association for the Advancement of Science, 1989; National Council of Teachers of Mathematics, 1989). These reforms call for radically new ways of conceiving of the teaching and learning of mathematics and science. They are relevant to my discussion of interdisciplinary education primarily because of the suggestions they make regarding the relations of these disciplines to problem solving and everyday experience. In short, these proposals stem from a general acceptance of the constructivist theory of learning and a growing unease with the sharp traditional distinction between theoretical understanding in the disciplines and practical activity. As Toulmin (1977) suggests, even the traditional disciplines, conceived as growing, changing social institutions attempting to solve persistent human problems, often oscillate back and forth between discipline-oriented phases and problem-oriented phases. Mathematics and science education, with the new proposed standards, may now be entering one of the problem-oriented phases, and in this respect will begin to appear more interdisciplinary.

The new standards proposed by the National Council of Teachers of Mathematics (NCTM) (1989) presuppose a radically different conception of mathematics than the traditional, algebra, geometry, trigonometry, calculus way of proceeding. Goals for students are much more process oriented. They include the ideas that students should learn to value mathematics, reason mathematically, communicate mathematically, become confident of their mathematical abilities, and become mathematical problem solvers. In addition to more traditional concepts such as whole numbers, algebra, and geometry, the standards involve problem solving; communication; reasoning; seeing connections; understanding measurement, statistics, and probability; seeing patterns and relations; and attaining mathematical power. The sample problems illustrating the standards also move beyond the artificial "story problems" so familiar to most of us to encompass the multistep solution of real problems found in society. In this way, mathematics itself becomes an interdisciplinary approach to solving certain kinds of social problems involving quantity, space, measurement, and statistics.

The American Association for the Advancement of Science (AAAS) takes a similar tack in its Project 2061 (the name is derived from the next return of Halley's Comet) report, *Science for All Americans* (1989). The very title of the report indicates its focus not simply on those who will become the next generation of scientists, but on a scientific understanding for all. AAAS argues that our increasingly complex world requires a certain level of scientific literacy of all of us. This is to be accomplished by softening the boundaries between the traditional areas, turning the "layer cake" of current science education on its side and, in so doing, emphasizing the connections among the traditional areas of biology, chemistry, physics, and earth science.

AAAS (1989) proposes to lessen the amount of detailed knowledge that students are expected to retain and to increase the understanding of the essential processes and presuppositions of science, including the place of science in the overall history of ideas. The social implications of science, mathematics, and technology are stressed, along with inculcating scientific habits of mind. These habits of mind include the internalization of the scientific values of the respect for and use of evidence and reasoning; informed beliefs about the social costs and benefits of science; a positive attitude toward being able to understand science and mathematics; computational skills, especially with regard to estimation of reasonable answers; manipulation and observational skills, including the use of a computer; communication skills, including the use of graphs, tables, and diagrams; and critical skills that will enable the student to evaluate arguments and claims that invoke the mantle of science.

As with the NCTM standards (1989), *Science for All Americans* (1989) focuses more on process and underlying conceptions and connections than on traditional disciplinary content. There is a clear commitment to demystifying science and making it accessible to all Americans. Turning the layer cake on its side in curriculum development will necessitate at least an initial level of interdisciplinary work, as teachers of biology, chemistry, and physics will have to study the other sciences to learn the connections and underlying principles that they will need to teach their students. The focus on science, technology, and society also provides a clear recognition of the places at which science as a set of disciplines must connect with other disciplines.

Both the NCTM (1989) and the AAAS (1989) projects are involved in exploring the ways in which teacher education must be changed in order to accommodate these new conceptions of mathematics and science. Another report has recently been issued on this topic, A *Call for Change: Recommendations for the Mathematical Preparation of Teachers of Mathematics* (Mathematical Association of America, 1991). Almost all of the discussion calls for a much more problem-focused and contextualized approach to the preparation of teachers. This is fully consonant with the problem-solving theme found in the interdisciplinary literature, while the calls to integrate mathematical and scientific concepts partake of the transdisciplinary approach. Hursh et al.'s (1983) discussion of the implementation of a skills-based, problem-solving general education curriculum discussed above is also relevant here. Their emphasis on the tentative nature of knowledge, intellectual skills, problem solving, the methodologies of problem solving, reflection on the material, and one's approach to it are all very consonant with the NCTM and AAAS rhetoric.

Both the NCTM (1989) and the AAAS (1989) reports stress the connections that must be made between theoretical understanding and practical activity, especially for those who may not go on to make a career of mathematics or science. Both reports are also clearly informed by the constructivist approach to learning with their emphasis on student activities as opposed to the simple delivery of content. Neither has much to say, however, about whether or not these bold new conceptions of mathematics and science actually are teachable to individuals or whether we may not have to approach mathematical and scientific literacy from a more social perspective. Of all the disciplines, mathematics and science have perhaps the best-developed and most rigorous standards for competence in the field. Whether one can, by focusing on processes, skills, problem solving, and the social and historical connections of science, escape the disciplinary paradox remains to be seen. It is likely that many "real" scientists will find such interdisciplinary approaches inadequate. At the same time, there is also afoot in the land a serious reexamination, discussed above, of the dominance of the traditional disciplines. In any event, such efforts have much to gain from close collaboration with those who have been toiling for some time in the interdisciplinary vineyards.

The coalition of essential schools

I turn now to Ted Sizer's Coalition of Essential Schools and its associated program with the Education Commission of the States, Re:Learning (Sizer, 1988). Sizer's Coalition of Essential Schools arises out of his Study of High Schools, documented in three volumes (Hampel, 1986; Powell, Farrar, & Cohen, 1985; Sizer, 1984). The Coalition is an ever-increasing group of schools committed to restructuring education around the common principles that emerged from the Study of High Schools. Re:Learning is an effort to link the restructuring effort going on at individual schools with district- and statewide educational reform. It is supported primarily by the Education Commission of the States, a nonprofit, nationwide, interstate compact formed to help governors, state legislatures, state education officials, and others develop policies to improve education.

- The so-called Common Principles of the Coalition (Coalition of Essential Schools, 1988) include the following:
- The school should focus on helping adolescents learn to use their minds well.
- Each student should master a limited number of essential skills and areas of knowledge. "Less is more."
- The school's goals should apply to all students.
- Teaching and learning should be personalized.
- The governing practical metaphor of the school should be student-as-worker and teacher-as-coach.
- The diploma should be awarded upon a successful final demonstration of mastery—an exhibition.
- The tone of the school should stress values of unanxious expectation.
- The principals and teachers should perceive of themselves as generalists first and disciplinary specialists second.
- Student loads should not exceed 80 per teacher and costs should not exceed 10% more than traditional schools.

Sizer and his colleagues believe that these principles, conscientiously implemented in schools, will have a revolutionary impact on schooling. The Coalition is continuing its efforts, although with a number of problems analogous to those of instituting interdisciplinary education efforts (Chion-Kenney, 1987; see the newsletters of the Coalition of Essential Schools, 1984-1991).

Several features of Coalition schools reflect the interdisciplinary emphasis inherent in the principles. First, the notions that less is more and that teachers should be generalists first and specialists second clearly reflect a movement toward interdisciplinarity. Students are supposed to master the material and make significant intellectual achievements rather than just grasp traditional content. The requirement for an "exhibition" of mastery (i.e., a kind of practical "doing") rather than the accumulation of a list of courses taken is another hallmark of interdisciplinary work. It is also clear that what is intended here is a synthesis and integration rather than simply a multidisciplinary gathering together of traditional disciplines, although it does not appear that the Coalition is hostile to the traditional disciplines or that it attempts to foster some new notion of transdisciplinarity.

The distinction between theoretical and practical knowledge seems, in Coalition schools, to tilt toward the primacy of the practical. This is seen in the deemphasis of traditional subjects and the central place of an exhibition as a demonstration of mastery. The exhibition is described as a demonstration that the student can do important things using the mind. Typical exhibitions appear to rely heavily on real-life projects such as environmental impact studies, artistic performances, and the like. They are not typically discipline-sized chunks, but rather real social problems.

The rhetoric for Coalition schools also seems to emphasize teamwork, although perhaps more so for faculty and staff than for students. It is clear that Sizer believes that the professionalization of teaching must result in much more interaction and sharing of knowledge among school professionals than is now the case. As generalists, teachers and other staff are also expected to take on multiple obligations such as teacher-counselor manager and must have a sense of commitment to the whole school.

Clearly, Coalition schools subscribe heavily to a version of the constructivist theory of learning. The student-as-worker exemplifies the spirit of the constructivist position, as does the emphasis on students using their minds well to accomplish important things, although there could be some concern about who decides what the "important things" are and on what basis. It may be that these important things are still to be taken largely from the received wisdom, and, therefore, they might not obviously represent a fresh look at what the real problems of society are.

Although the rhetoric of interdisciplinary education is somewhat muted in the Coalition of Essential Schools, it is, nevertheless, clearly present, at least as a means to the end of restructuring schooling. As such, it is also predictable that many of the typical problems of interdisciplinary education also bedevil the work of the Coalition. In particular, several versions of the disciplinary paradox seem to attend the development of Coalition schools (Sizer, 1989; see the newsletters of the Coalition of Essential Schools, 1984-1991). As might be expected, teachers in Coalition schools have a good deal of trouble seeing themselves as generalists. Their bases in the disciplines provide a sense of security, and venturing forth from those secure bases is problematic. Similarly, there is a good deal of skepticism regarding the actual worthwhileness of work in a Coalition school. Parents, administrators, and policymakers want to know how the students will do on standardized tests, yet if the emphasis is on depth instead of coverage and on essential skills instead of the traditional academic disciplines, it is not at all clear how well the students will do on traditional, discipline-based assessment procedures.

It is at this point that the emphasis on the exhibition becomes particularly important. Coalition proponents realize that as one attempts to move away from the disciplines, new standards and forms of evaluation must be developed, or else the perspectives of the disciplines will devalue the new work. Coalition schools are beginning to join forces with those who are advocating the development of new forms of authentic assessments (Gardner, in press; Lave, 1988; Resnick & Resnick, in press; Stiggins, 1988; Wiggins, 1989; Wolf, Bixby, Glenn, & Gardner, 1991), and will need to pay attention to the extent to which the exhibitions really do appear to address real problems and issues of society. Recall that one way out of the disciplinary paradox was to judge interdisciplinary work on its success in dealing with practical problems. On the positive side the communities of inquiry, composed of both students and faculty, in Coalition schools will almost surely help each of the participants to experience the extent to which knowledge is a social construction. Work in the Coalition would certainly benefit from a more explicit acknowledgment that much of what is being advocated really is interdisciplinary education, with all of the attendant problems and challenges.

"Learning in school and out"

Last, let me turn to Lauren Resnick's AERA presidential address, "Learning in School and Out" (1987). In this address Resnick draws four sharp contrasts between learning in schools and learning outside of schools, and, although her purpose was not to address directly the issue of interdisciplinary education, the distinctions she draws are clearly relevant.

Resnick (1987) first contrasts the kind of individual cognition we value in school with the shared cognition valued outside of schools. It seems that our social systems embody the realization that we cannot, in very many instances, solve the problems we must solve if we rely solely on individual performances. Yet, in school, we insist that the performances be individual. This insistence may be compatible with the function of the school as a social and economic sorting mechanism that assigns people to scarce desirable positions. However, it is not clear that it is helpful educationally in promoting student learning. Why do we not use socially cooperative modes of instruction and learning instead of individually competitive modes? Real work, even in the traditional disciplines, more often than not takes place collaboratively.

The second contrast cited by Resnick (1987) is the difference between pure mentation in school subjects and tool manipulation outside (see also Lave, 1988). Why, for example, do we continue to insist that students learn the multiplication tables instead of allowing them to use calculators as tools to obtain a deeper knowledge of mathematics? This contrast echoes the tendency for the traditional disciplines to focus on theoretical understanding and accepted ways of doing things, while interdisciplinary problems require practical reasoning and solutions.

The difference between symbol manipulation in school and highly contextualized reasoning outside of school forms Resnick's (1987) third contrast (see also Lave, 1988). She cites an interesting example of the difference between symbol manipulation in solving a mathematics problem and contextualized reasoning. The problem involves deciding how much more money would be needed to buy an ice cream cone costing 60 cents if one had in hand a quarter, a dime, and two pennies. The standard school solution involves the calculation that 23 cents more is needed. The typical situated real-world solution involves looking for some additional coins, perhaps another quarter. As interdisciplinarians point out, real problems seldom come in discipline-shaped chunks (Roy, 1979), and this helps explain why doing well in school often has so little effect on how well one does in the real world. The reasoning is simply different in the different places, and at least one impetus for interdisciplinary work, the solution of real problems, tends to focus on the situated reasoning needed in the real world.

Resnick's (1987) fourth contrast is that between generalized learning in school and situation-specific competencies outside. A version of the disciplinary paradox can be constructed here as well. On the one hand, generalized theoretical learning does not always apply directly to or translate well into specific situations. Thus, we often need some sort of situation-specific problem solving. On the other hand, if one only learns to solve specific problems, then the adaptability to new situations is suspect. As I have been arguing, interdisciplinary education provides a way of beginning to see how to relate theoretical knowledge to practical knowledge.

Resnick's (1987) discussion of learning in school and out raises quite forcefully several of the conceptual themes I have used to discuss interdisciplinary education. First, the distinction between theoretical and practical wisdom is implicit in all of Resnick's contrasts. Schools, especially secondary schools, in their massive reliance on traditional disciplines, clearly emphasize theoretical wisdom. The real world, however, often demands practical competence. Second, Resnick's powerful descriptions of the social nature of learning and performance outside of schools call into serious question whether or not interdisciplinarity can be reasonably located within individuals rather than in social systems. Finally, the idea of situation-specific learning implies the need for much more work on just how we do, indeed, seem to construct meaning from our individual experiences and yet, nevertheless, are able to deal with new situations that do not differ too radically from the ones in which we have learned the skills and competencies. The traditional answer is that we learn general principles and then apply them to specific situations. The interdisciplinary answer seems to be that we must bring different perspectives to bear in learning how to see specific situations in new and useful ways.

Conclusion

The impetus to interdisciplinary education comes from a variety of different directions. One impetus is integrative in the epistemological sense. The disciplines appear to have fragmented knowledge, and it would be better to have a unified system of knowledge. This impetus results in attempts to develop transdisciplinary approaches to knowledge, in many of the efforts to devise some kind of core curriculum, and perhaps in the new standards proposed for mathematics and science teaching. The underlying idea seems to be that somehow we should be able to unify knowledge in the head of the individual. Typically, this impetus is directed primarily at theoretical understanding, and is particularly subject to the disciplinary paradox. How can we develop standards of evaluation of a unified core curriculum separate from the standards of the disciplines?

Another impetus is the need to solve the practical problems of society. These problems seldom come in discipline-shaped chunks. At issue here is the extent to which these problems are defined by society as it exists as opposed to being the problems that a restructured society ought to address if the deficiencies of the disciplines were corrected. The press for interdisciplinarity can be conceived both as inherently conservative of how we currently understand our human milieu or as progressive in moving us forward and out of the straitjackets of the disciplines. Interdisciplinary research on weapons delivery systems may be an example of the former, while interdisciplinary research in women's studies may exemplify the latter.

Another press that seems to combine elements of both the theoretical and practical comes from the imperatives of civic and cultural education. Our educational system has always had as one of its chief functions the transmission of the culture. This culture has both theoretical and practical aspects. Indeed, as I have noted, one of the more interesting current debates has to do with the extent to which we do have or ought to have a common culture in the United States or whether we have moved or ought to move to a more multicultural perspective. One of the strengths of our liberal democratic heritage, in its emphasis on tolerance and the marketplace of ideas, is that the possibility of embracing a multicultural perspective is not an incoherent idea. In any case, interdisciplinary education is central to the debate in that it provides a major stimulus for the dialogue.

Indeed, the social nature of much interdisciplinary thought and education may even suggest a practical solution to the problem of possible social fragmentation that may follow upon an overly enthusiastic embracing of multiculturalism. We need only recall Campbell's (1969) fish-scale model and seek out those individuals who, though still specialized at the fringes of an existing culture, overlap with other cultures. We must then expand our mainstream cultures to incorporate these individuals into social arrangements in which we can take advantage of their bridging skills. Perhaps we can, through new social arrangements, celebrate our diversity and see it as a strength.

Finally, the impetus for interdisciplinary thought probably also comes from the disciplines themselves, at least if they are vibrant and active. As long as the disciplines remember their fundamental roles as systematized ways of helping human beings deal with the major problems of being human, they will never for too long be abstract and disconnected from practical affairs. Interdisciplinary problem solving and theoretical reflection will serve to remind the disciplines that they must constantly anchor their work in the real world of human thought and activity.

The idea of interdisciplinary teaching and learning is a powerful and appealing one. We must, however, think hard about its limitations. Can we overcome the disciplinary paradox in its various manifestations? Can interdisciplinary knowledge be integrated in the heads of individuals, or must it be located in social groupings? Does multiculturalism constitute a kind of locating of interdisciplinary knowledge in groups? What are the relations between theoretical understanding and practical wisdom? If we address these issues carefully and systematically, perhaps the opportunities promised by interdisciplinary education will not be insurmountable after all.

References

American Association for the Advancement of Science. (1989). *Science for all Americans: Project 2061*. Washington, DC: Author.

Anderson, R. C., & Pearson, P. D. (1984). A schema-theoretic view of basic processes in reading comprehension. In P. D. Pearson (Ed.), *Handbook of reading research* (pp. 255-292). New York: Longman.

Aristotle. (1941). Nicomachean ethics, Book VI. In R. McKeon (Ed.), *The basic works of Aristotle*. New York: Random House.

Association of American Colleges. (1991). *Liberal learning and arts and sciences majors: The challenge of connecting learning, Vol. 1*. Washington, DC: Author.

Association of American Colleges and Society for Values in Higher Education. (1990). Interdisciplinary resources. *Issues in Integrative Studies, 8*.

Bateson, G. (1979). *Mind and nature: A necessary unity*. New York: Dutton.

Becher, A. (1989). *Academic tribes and territories: Intellectual enquiry and cultures of the disciplines*. Milton Keynes, Australia: Open University.

Becher, A. (1990). The counter-culture of specialisation. *European Journal of Education, 25*, 333-346.

Benson, G. D. (1989). Epistemology and science curriculum. *Journal of Curriculum Studies, 21*, 329-344.

Berliner, D. (1986). In pursuit of the expert pedagogue. *Educational Researcher, 15*(7), 5-13.

Berliner, D. (1989). Implications of studies of expertise in pedagogy for teacher education and evaluation. In *New directions for teacher assessment: Proceedings of the 1988 ETS International Conference* (pp. 39-78). Princeton, NJ: Educational Testing Service.

Birnbaum, N. (1969, July—August). The arbitrary disciplines. *Change, the Magazine of Higher Learning*, pp. 10-21.

Boulding, K. (1956). *The image: Knowledge in life and society*. Ann Arbor: University of Michigan.

Boyer, E. (Ed.). (1981). *Common learning: A Carnegie colloquium on general education*. Washington, DC: Carnegie Foundation for the Advancement of Teaching.

Boyer, E. (1987). *The undergraduate experience in America*. NY: Harper & Row.

Boyer, E. (1990). *Scholarship reconsidered: Priorities of the professoriate*. Washington, DC: Carnegie Foundation for the Advancement of Teaching.

Brantlinger, P. (1990). *Crusoe's footprints: Cultural studies in Britain and America*. New York: Routledge.

Brown, J., Collins, D., & Duguid, P. (1989). Situated cognition and the culture of learning. *Educational Researcher, 18(1),* 32-42.

Caldwell, L. (1983). Environmental studies: Discipline or metadiscipline? *Environmental Professional,* 5, 247-258.

Campbell, D. (1969). Ethnocentrism of disciplines and the fish-scale model of omniscience. In M. Sherif & C. Sherif (Ed.), *Interdisciplinary relationships in the social sciences* (pp. 328-348). Chicago: Aldine.

Casey, B. (1986). The quiet revolution: The transformation and reintegration of the humanities. *Issues in Integrative Studies, 4,* 71-92.

Chion-Kenney, L. (1987, Winter). A report from the field: The Coalition of Essential Schools. *American Educator,* pp. 18-28.

Chubin, D., Porter, A., & Rossini, F. (1986). Interdisciplinary research: The why and the how. In D. Chubin, A. Porter, F. Rossini, & T. Connolly (Eds.), *Interdisciplinary analysis and research: Theory and practice of problem-focused research and development* (pp. 3-10). Mt. Airy, MD: Lomond.

Chubin, D., Porter, A., Rossini, F., & Connolly, T. (Eds.). (1986). *Interdisciplinary analysis and research: Theory and practice of problem-focused research and development.* Mt. Airy, MD: Lomond.

Clark, M., & Wawrytko, S. (Eds.). (1990). *Rethinking the curriculum: Toward an integrated, interdisciplinary college education.* New York: Greenwood.

Clifford, G., & Guthrie, J. (1988). *Ed school.* Chicago: U. of Chicago Press.

Clifford, J., & Marcus, G. (Eds.). (1986). *Writing culture: The poetics and politics of ethnography.* Berkeley: University of California.

Coalition of Essential Schools. (1984). *Prospectus.* Providence, RI: Coalition of Essential Schools, Brown University.

Coalition of Essential Schools. (1988). *The common principles of the Coalition of Essential Schools.* Providence, RI: Coalition of Essential Schools, Brown U.

Cochran-Smith, M., & Lytle, S. (1990). Research on teaching and teacher research: Issues that divide. *Educational Researcher,* 19(2), 2-11.

Commissioner's Task Force on Minorities: Equity and Excellence. (1989). *A curriculum of inclusion.* Albany, NY: State Education Department.

D'Souza, D. (1991). *Illiberal education: The politics of race and sex on campus.* New York: Free Press.

Dewey, J. (1916). *Democracy and education.* New York: Macmillan.

Dewey, J. (1933). *How we think.* New York: D. C. Heath.

Dewey, J. (1938). *Experience and education.* New York: Macmillan.

Eisner, E. (1981). On the differences between scientific and artistic approaches to qualitative research. *Educational Researcher, 10(4),* 5-9.

Elam, D. (1990). Ms. en abyme [Feminism in feminist scholarship]. *Social Epistemology, 4,* 293-308.

Elman, S., & Smock, S. (1985). *Professional service and faculty rewards: Toward an integrated structure.* Washington, DC: National Association of State Universities and Land-Grant Colleges.

Firestone, W. (1987). Meaning in method: The rhetoric of qualitative and quantitative research. *Educational Researcher, 16(7),* 16-21.

Fish, S. (1989). Being interdisciplinary is so very hard to do. *Profession 89* (pp. 15-22). New York: Modern Languages Association.

Gaff, J. (1989). The resurgence of interdisciplinary studies. *National Forum,* 69(2), 4-5.

Gardner, H. (1983). *Frames of mind.* England: Cambridge University Press.

Gardner, H. (in press). Assessment in context: The alternative to standardized testing. In B. Gifford, M. O'Connor, & M. Catherine (Eds.), *Rethinking aptitude, achievement, and assessment.* Boston: Kluwer Academic Publ.

Geertz, C. (1980). Blurred genres. *American Scholar, 42,* 165-179.

General education. (1989, July/August). *Change, The Magazine of Higher Learning.*

Giroux, H., & McLaren, P. (1986). Teacher education and the politics of engagement: The case for democratic schooling. *Harvard Educational Review,* 56, 213-238.

Goodlad, J. (1983). *A place called school.* New York: McGraw-Hill.

Group for Research and Innovation. (1975). *Interdisciplinarity: A report by the Group for Research and Innovation.* Regents Park, England: Group for Research and Innovation, The Nuffield Foundation.

Guba, E., & Lincoln, Y. (1988). Do inquiry paradigms imply inquiry methodologies? In D. Fetterman (Ed.), *Qualitative approaches to evaluation in education: The silent scientific revolution* (pp. 89-115). New York: Praeger.

Hamlyn, D. W. (1978). *Experience and the growth of understanding.* London: Routledge & Kegan Paul.

Hampel, R. (1986). *The last little citadel.* Boston: Houghton Mifflin.

Hirsch, E. D. (1987). *Cultural literacy: What every American needs to know.* Boston: Houghton Mifflin.

Hirst, P. H. (1974). *Knowledge and the curriculum: A collection of philosophical papers.* London: Routledge & Kegan Paul.

Holmes Group. (1986). *Tomorrow's teachers.* East Lansing, MI: Author.

Holmes Group. (1990). *Tomorrow's schools.* East Lansing, MI: Author.

Howe, K. (1988). Against the quantitative-qualitative incompatibility thesis (or, dogmas die hard). *Educational Researcher,* 17(8), 10-16.

Howe, K., & Eisenhart, M. (1990). Standards for qualitative and quantitative research: A prolegomenon. *Educational Researcher, 19(4),* 2-9.

Hursh, B., Haas, P., & Moore, M. (1983). An interdisciplinary model to implement general education. *Journal of Higher Education, 54,* 42-59.

Interdisciplinary studies. (1978, August). *Change, The Magazine of Higher Learning,* pp. 6-48.

Jacob, E. (1988). Clarifying qualitative research: A focus on traditions. *Educational Researcher, 17(1),* 16-24.

Judge, H. (1982). *American graduate schools of education: A view from abroad.* New York: Ford Foundation.

Kant, I. (1961). *Critique of pure reason, translated by Norman Kemp Smith.* London: Macmillan.

Kennedy, M. (1987). *Inexact sciences: Professional education and the development of expertise* (Issue Paper 87-2). East Lansing: National Center for Research on Teacher Education, Michigan State University.

Kirst, M. (1991). Improving children's services: Overcoming barriers, creating new opportunities. *Phi Delta Kappan, 72,* 615-618.

Klein, J. T. (1985). The interdisciplinary concept: Past, present, and future. In L. Levin & I. Lind (Eds.), *Interdisciplinarity revisited: Re-assessing the concept in the light of institutional experience* (pp. 104-136). Stockholm: OECD/CERI, Swedish National Board of Universities and Colleges, Linköping University.

Klein, J. T. (1990a). *Interdisciplinarity: History, theory, and practice.* Detroit: Wayne State University Press.

Klein, J. T. (1990b). Interdisciplinary resources: A bibliographical reflection. *Issues in Integrative Studies: Interdisciplinary Resources, 8,* 35-67.

Kockelmans, J. (Ed.). (1979). *Interdisciplinarity and higher education.* University Park: Pennsylvania State University Press.

Kolodny, A. (1984). *The land before her: Fantasy and experience of the American frontier, 1630-1860.* Chapel Hill: University of North Carolina.

Kroker, A. (1980). Migration across the disciplines. *Journal of Canadian Studies, 15,* 3-10.

Kuhn, T. (1970). *The structure of scientific revolutions* (enlarged ed.). Chicago: University of Chicago Press.

Kuhn, T. (1974). Second thoughts on paradigms. In F. Suppe (Ed.), *The structure of scientific theories* (pp. 459-482). Urbana: University of Illinois.

Landau, M., Proshansky, H., & Ittelson, W. (1962). The interdisciplinary approach and the concept of the behavioral sciences. In N. Washburne (Ed.), *Decision, values and groups, II* (pp. 7-25). New York: Pergamon.

Lave, J. (1988). *Cognition in practice: Mind, mathematics, and culture in everyday life*. New York: Cambridge University Press.

Levin, L., & Lind, I. (Eds.). (1985). *Interdisciplinarity revisited: Re-assessing the concept in the light of institutional experience*. Stockholm: OECD, Swedish National Board of Universities and Colleges, Linköping University.

Martin, J. (1982). Two dogmas of curriculum. *Synthese, 51,* 5-20.

Mathematical Association of America. (1991). *A call for change: Recommendations for the mathematical preparation of teachers of mathematics*. Washington, DC: Author.

McPeck, J. (1981). *Critical thinking and education*. New York: St. Martin's. Miller, R. (1982). Varieties of interdisciplinary approaches in the social sciences. *Issues in Integrative Studies, 1,* 1-37.

National Council of Teachers of Mathematics. (1989). *Curriculum and evaluation standards for school mathematics*. Washington, DC: Author.

Newell, W. (Ed.). (1986). *Interdisciplinary undergraduate programs: A directory*. Oxford, OH: Association for Integrative Studies.

Newell, W. (1988). Interdisciplinary studies are alive and well. *Association for Integrative Studies Newsletter, 10(1),* 6-8.

Organization for Economic Cooperation and Development. (1972). *Interdisciplinarity: Problems of teaching and research in universities*. Paris: Author.

Perry, W. (1968). *Forms of intellectual and ethical development in the college years*. New York: Holt, Rinehart, and Winston.

Perry, W. (1981). Cognitive and ethical growth: The making of meaning. In A. Chickering (Ed.), *The modern American college* (pp. 76-116). San Francisco: Jossey-Bass.

Peshkin, A. (1988). In search of subjectivity-One's own. *Educational Researcher, 17(7),* 17-22.

Peters, T., & Waterman, R. (1982). *In search of excellence: Lessons from America's best-run companies*. New York: Warner Books.

Petrie, H. G. (1976). Do you see what I see? The epistemology of interdisciplinary inquiry. *Educational Researcher, 5(2),* 9-15.

Petrie, H. (1981). *The dilemma of enquiry and learning*. Chicago: University of Chicago Press.

Petrie, H. (1987). Teacher education, the liberal arts, and extended preparation programs. *Educational Policy, 1(1),* 29-42.

Phillips, D. C. (1983). After the wake: Postpositivistic educational thought. *Educational Researcher, 12(5),* 4-12.

Phillips, D. C. (1987a). *Philosophy, science, and social inquiry*. NY: Pergamon.

Phillips, D. C. (1987b). Validity in qualitative research: Why the worry with warrant will not wane. *Education and Urban Society, 20,* 9-24.

Phillips, D. C., & Soltis, J. F. (1985). *Perspectives on learning.* NY: Teachers College Press.

Piaget, J., & Inhelder, B. (1969). *The psychology of the child.* NY: Basic Books.

Porter, A., & Rossini, F. (1984). Interdisciplinary research redefined: Multi-skill, problem-focused research in the STRAP framework. *R & D Management, 14,* 105-111.

Powell, A., Farrar, E., & Cohen, D. (1985). *The shopping mall high school.* Boston: Houghton Mifflin.

Ravitch, D., & Finn, C. (1987). *What do our 17-year-olds know? A report on the first national assessment of history and literature.* New York: Harper & Row.

Resnick, L. (1987). Learning in school and out. *Educational Researcher, 16*(9), 13-20.

Resnick, L., & Resnick, D. (in press). Assessing the thinking curriculum. In B. Gifford, M. O'Connor, & M. Catherine (Eds.), *Rethinking aptitude, achievement, and assessment in testing.* Boston: Kluwer Academic Publ.

Rich, D., & Warren, R. (1980). The intellectual future of urban affairs: Theoretical, normative, and organizational options. *Social Science Journal, 17*(2), 53-66.

Robbins, B. (1987). Poaching off the disciplines. *Raritan, 6*(4), 81-96.

Roy, R. (1979). Interdisciplinary science on campus: The elusive dream. In J. Kockelmans (Ed.), *Interdisciplinarity and higher education* (pp. 161-196). University Park: Pennsylvania State University Press.

Schön, D. (1983). *The reflective practitioner.* New York: Basic Books.

Schön, D. (1987). *Educating the reflective practitioner.* San Francisco: Jossey-Bass.

Shulman, J., & Colbert, J. (1987). *The mentor teacher casebook.* San Francisco: Far West Laboratory for Educational Research and Development.

Shulman, J., & Colbert, J. (1988). *The intern teacher casebook.* San Francisco: Far West Laboratory for Educational Research and Development.

Shulman, L. (1986). Those who understand: Knowledge growth in teaching. *Educational Researcher, 15*(2), 4-14.

Shulman, L. (1987a). Assessment for teaching: An initiative for the profession. *Phi Delta Kappan, 69,* 38-44.

Shulman, L. (1987b). Knowledge and teaching: Foundations of the new reform. *Harvard Educational Review, 57,* 1-22.

Shulman, L. (1988). Disciplines of inquiry in education: An overview. In R. Jaeger (Ed.), *Complementary methods for research in education* (pp. 3-17). Washington, DC: American Educational Research Association.

Simons, H. (Ed.). (1989). *Rhetoric in the human sciences.* London: Sage.

Simons, H. (Ed.). (1990). *The rhetorical turn: Invention and persuasion in the conduct of inquiry.* Chicago: University of Chicago Press.

Sizer, T. (1984). *Horace's compromise: The dilemma of the American high school.* Boston: Houghton Mifflin.

Sizer, T. (1988, August). Creating a society that thinks: Re:Learning. *State Government News,* pp. 20-21.

Sizer, T. (1989). Diverse practice, shared ideas: The essential school. In H. Walberg & J. Lane (Eds.), *Organizing for learning: Toward the 21st century* (pp. 1-8). Reston, VA: National Association of Secondary School Principals.

Social Studies Syllabus Review and Development Committee. (1991). *One nation, many peoples: A declaration of cultural independence.* Albany, NY: State Education Department.

Sternberg, R. (1985). *Beyond IQ: A triarchic theory of human intelligence.* Cambridge, England: Cambridge University Press.

Stiggins, R. (1988). Revitalizing classroom assessment. *Phi Delta Kappan, 69,* 363-368.

Stimpson, C. (1988). *Where the meanings are.* New York: Methuen.

Stimpson, C., & Cobb, N. (1986). *Women's studies in the United States.* New York: Ford Foundation.

Taylor, A. M. (1969). Integrative principles and the educational process. *Main Currents in Modern Thought, 25,* 126-133.

Toulmin, S. (1972). *Human understanding: Vol. 1.* Princeton, NJ: Princeton University Press.

Toulmin, S. (1977). From form to function: Philosophy and history of science in the 1950s and now. *Daedalus, 1,* 143-162.

Toulmin, S. (1982). *The return to cosmology: Postmodern science and the theology of nature.* Berkeley: University of California Press.

Trow, M. (1984/1985). Interdisciplinary studies as a counterculture: Problems of birth, growth, and survival. *Issues in Integrative Studies, 4,* 1-15.

Turner, B. (1990). The interdisciplinary curriculum: From social medicine to postmodernism. *Sociology of Health and Illness, 12(1),* 1-23.

Undergraduate College. (1990). *A new general education curriculum for arts and sciences students at UB: A proposal from the undergraduate college to the university (revised).* Buffalo: State University of New York at Buffalo.

Wiggins, G. (1989). A true test: Toward more authentic and equitable assessment. *Phi Delta Kappan, 70,* 703-713.

Wolf, D., Bixby, J., Glenn, J. III, & Gardner, H. (1991). To use their minds well: Investigating new forms of student assessment. In G. Grant (Ed.), *Review of research in education* (Vol. 17, pp. 31-74). Washington, DC: American Educational Research Association.

[1992]
Knowledge, Practice, and Judgment

Introduction

The traditional conception of the relationship between research and practice is that we somehow "apply" the knowledge gained from research to practice. With respect to teachers' knowledge this presupposition is enshrined in teacher education through the routine of presenting prospective teachers with research-based theories of learning and instruction and then giving them experience in applying this knowledge to classroom situations during student teaching. This model also pervades the in-service workshops so prevalent during a teacher's career.

I want to argue that except in the most routine of situations, the conception of teachers "applying" knowledge to practice is fundamentally flawed. This suggests that if we persist in the notion of teachers applying knowledge to practice, we are largely committing ourselves, albeit unwittingly, to a conception of teaching as routine, technical, and susceptible to top-down micro-management. On the other hand, if we can provide a better conception of the relationship between teacher knowledge and teacher action, we may be able to defend a more professional vision of teaching.

My objectives are four-fold. First, I will demonstrate that the conception of applying knowledge to practice is adequate only in the most routine situations. Second, I will urge that the currently popular and promising conception of the teacher as reflective practitioner ultimately requires a notion of professional judgment which joins both thought and action. Third, I will sketch a preliminary analysis of what professional judgment is. Finally, I will have some suggestions as to how a properly explicated notion of professional judgment can contribute to understanding the kind of accountability appropriate to a true profession of teaching.

First published in: Educational Foundations, Volume 6, Number 1, Winter 1992. pp. 35-48

Professional expertise

Mary Kennedy (1987) has identified a number of ways in which we might conceive of the relationship between knowledge and action, theory and practice. One of these ways, the technical skills version of what Kennedy calls "professional expertise," does, indeed, consist of "applying" a kind of knowledge to action. This kind of knowledge involves a notion of "technical skills" which can be readily identified and taught to would-be practitioners and which, if utilized, have some sort of clear relationship to improved practice. There are skills, like planning lessons, waiting after asking questions, calling upon silent members of a class, and the like, which clearly are appropriate things teachers should do and can be taught to do. Furthermore, such skills are typically applied in fairly routine, straightforward situations.

The problem, as Kennedy points out, is that an overemphasis on the identification and acquisition of skills often ignores the issue of the rationale for their use. If teachers' work were highly structured and routine, say, like an assembly line, then it might be sufficient to make sure that teachers simply acquired the requisite skills. The rationale for the enterprise would be found in its overall organization or in the detailed supervision of the "assembly line" by more knowledgeable persons, perhaps principals or curriculum specialists. Teachers would only need to apply their skills to absolutely predictable and regular classroom situations. Unfortunately, as we all know, classrooms are nothing like assembly lines, despite the attempts of some to make them so.

Kennedy's second conception of expertise, the application of theory or general principles, would seem to remedy the problem of focusing solely on the nature and acquisition of the specific technical skills. If one learns not only the technical skills, but also the rationale and general principles underlying the skills, then one will be able to apply the skills to situations more diverse than an absolutely predictable assembly line. The conception here, one which is largely, if implicitly, followed in teacher preparation programs, is that as one comes to understand the principles and general theory underlying the use of technical skills, one will be able to adapt one's behavior to changing circumstances.

There are, however, several problems with this view of expertise as well. First, real situations do not present themselves as obviously identifiable cases of general principles. One of the key requirements for applying general principles is that a kind of perceptual learning take place so that the teacher can learn to see this situation as a case of that principle and this other situation as a case of a different principle (Kennedy, 1987; Petrie, 1976, 1981). This observation lends credence to the continuing strong support for a "case" methodology in preparing teachers. Indeed, it may be that what education needs most of all are exemplary cases of the central concepts of teaching. One does not "apply" knowledge to neutral cases, rather one structures experiential situations in terms of certain concepts exemplified by paradigmatic cases.

Second, however, even if one recognizes a particular situation as a case of a general principle, the problem is that the situation is particular and the principle is general. How does one come, on this view, to be able to adjust the general principle to the particularities of practice? We are all familiar with teachers who know their theory and who can even recognize a situation as calling for a particular principle and yet be totally unable to adjust the principle to the situation. For example, it is not at all unusual to observe a teacher who knows that one ought not reward disruptive student behavior, who recognizes that a given student is being disruptive, but who is unable to do anything other than scream at the student, thereby providing precisely the attention the student craves. At the same time other teachers are marvelously adaptive in avoiding the situations which lead to the disruption or changing the focus of attention if the disruption occurs. What explains the difference?

One of the problems in accounting for those who can and those who cannot adapt to varying situations may lie in our not fully realizing that real-life situations can often be seen as falling under a multiplicity of propositionally formulated principles. The question then arises as to which principles we should apply, or how we should weight or modify them. Is it that we simply need to identify higher order principles and also teach these to our would-be teachers? These second-order principles would also be propositional in nature and would, on this conception be taught as a kind of higher order theory, which would then be applied to cases of choosing among lower order theoretical principles in a given situation. But how would the decisions be made on "applying" these higher order principles?

Indeed, we are caught here in a kind of logical infinite regress. Whenever we have trouble "applying" principles stated as propositional knowledge to practical situations, we postulate a knowledge that tells us how to make the application. But how do we know when and in what situations to apply this new knowledge of applications? Do we need a higher order knowledge connecting the propositional knowledge of how to apply knowledge to the propositional knowledge which we wish to apply? Could the process ever end? It would seem that the notion of "applying" knowledge may be at the root of our problems here (Petrie, 1981). We really only apply knowledge in the most routine of situations, situations for which a recipe can be constructed and which do not vary much at all from case to case.

We can speak of "applying" knowledge to situations like cooking and theorem proving because the situations are all reasonably well-defined, easily recognized, and fall under only a few principles. However, as soon as the situation becomes even a little bit complex, as, for example, when we try to formulate "important" theorems to prove, adaptations must be made. Even cooks must sometimes adapt their recipes to unusual circumstances. Perhaps they only have larger eggs (more liquid than the recipe calls for) or a flour which absorbs more liquid than usual. Human beings are able to vary their behavior constantly in order to do the same thing time after time, e.g., keep the class on track.[2] The notion of "applying" knowledge to practice completely fails to capture our ability to behave in constantly varying ways which bring the situation under the general principles which are guiding our actions.

Kennedy's third conception of expertise is that of "critical analysis." The critical analysis conception of professional expertise is best exemplified by legal education where the goal is to get the students to "think like lawyers." Essentially, such an approach takes very seriously the notion of bringing people to see the situations they encounter in terms of the principles of the field. It is little wonder, then, that the case method is so popular in legal education. As I noted above, case knowledge is precisely devoted to getting students to "see" situations as falling and failing to fall under key concepts and principles.

A notion of expertise which focuses almost exclusively on how to perceive and understand the situations one encounters in the appropriate terms is what Kennedy has in mind when she speaks of "critical analysis." However, there is a distinct possibility that not all situations are most felicitously analyzed from just one view point, no matter how powerful. As the old saw has it, once one has a hammer, then it is all too easy to see the whole world as something to be hammered, even if that is inappropriate. Furthermore, the critical analysis approach downplays the transition from understanding the problem in a certain way to acting on that way of seeing the situation. The link between thought and action is often ignored.

Kennedy's fourth notion of expertise is what she calls "deliberate action," taking the term from Schwab (1978). The idea is that of a reflective practitioner who both analyzes the given situation and acts within it. This approach gives promise of a direct attack on the problem of just how to relate thought and action, although the exact nature of the connection remains unclear. It also presupposes that there is a constant interplay between the ways in which we understand the problems and the means we contemplate for dealing with those problems. In short, means and ends are in constant interaction, both in actual experiments and in thought experiments as one tries to deal with a constantly changing world. This approach would seem to have some potential for providing a truly illuminating account of the relation between teacher knowledge and teacher action.

Although less widely discussed than his popular notion of pedagogical content knowledge, Lee Shulman (1986, 1987a, 1987b) has also raised the question of just how teachers possess and use their knowledge. His discussion of what he calls "strategic knowledge" is relevant to the question of the relationship between thought and action. Strategic knowledge for Shulman is that knowledge which is used to decide what to do in particular cases. It is used when principles collide, when a situation can be seen as a case of x or a case of y and we need to decide how to treat it. In short, strategic knowledge for teachers is that which enables them to make the myriad non-trivial decisions called for each day regarding the actual conduct of teaching. It is the knowledge which would allow a teacher to go beyond the "applying general principles" conception of expertise identified by Kennedy. Furthermore, Shulman's well-known emphasis on teacher reflection suggests a way of fleshing out the "deliberate" in Kennedy's notion of deliberate action. Deliberate action would be that in which we consciously deliberate or think about what to do using all of the knowledge we have.

However, in order to account for wise decision-making, strategic knowledge must be of a different order than theoretical knowledge, conscious deliberation, or understanding. Theoretical knowledge is still propositional knowledge. Strategic knowledge, however, must be logically connected to decision and action. It cannot be just another proposition about decision and action. Shulman senses this difference in a revealing footnote. He says (1986, p. 14):

> It may well be that what I am calling strategic knowledge in this paper is not knowledge in the same sense as propositional and case knowledge. Strategic "knowing" or judgment may simply be a process of analysis, of comparing and contrasting principles, cases, and their implications for practice.

The suggested move away from structures of knowledge to the processes of knowing is exactly right as a way of understanding the relation between thought and action.

In his two influential books, *The Reflective Practitioner: How Professionals Think in Action* (1983) and *Educating the Reflective Practitioner* (1987), Donald Schon provides perhaps the best current description of what a process of knowing would have to be like in order to relate thought and action. Schon's conception of reflective practice forges an indissoluble bond between thought and action in the practice of professionals, precisely the kind of bond I have urged must be found if we are to avoid the problems I have noted above.

Reflective practice for Schon includes what he calls "knowing-in-action," "reflection-in-action," and "reflection on reflection-in-action." Knowing-inaction is the intelligence actually revealed in competent professional performance, as in a superb musical rendition, or a brilliantly delivered lecture. Only rarely can knowing-in-action be described discursively. As close as we could come to a discursive description would perhaps be the musical score or the lesson plan for the lecture. Reflection-in-action involves the ability to change course during some complex performance in response to changing and unanticipated circumstances. Jazz improvisation is one example, reshaping a lecture in response to a student's question would be another. Reflection on reflection-in-action is when we actually stop and try to describe knowing-in-action and reflection-in-action discursively. It can be done only by someone who already knows pretty much how to perform at a reasonable level of skill. Unlike Shulman, Schon seems to believe that we do not often reflect in order to practice more effectively, but rather we must have reached some level of competent practice in order to profitably reflect.

This leads Schon to describe the situation in professional practice as analogous to Plato's *Meno* paradox (See also Petrie, 1981). How can we ever learn anything new, for, if we do not already know what we are seeking, how could we recognize it when we learn it? On the other hand, if we already know what we are to learn, what is the point of learning?

The answer to the dilemma for Schon is to put an emphasis on the professional practicum as the centerpiece of a curriculum to educate the reflective practitioner. The notion of cases as paradigm examples of good practice fits neatly into the practicum. In the beginning the student does not know what he or she is to learn. Nor could the teacher explain what is to be learned in any straightforward sense. However, through a variety of coaching strategies, involving joint practical problem-solving, occasionally an insistence by the coach that the student simply do as the coach does, and joint reflection by student and coach on the process, amazingly most students become independent practitioners. They are able not only to see the situations in generally the same ways as other mature professionals do, Kennedy's conception of critical analysis, but also to perform as a professional does, not in slavish imitation of their mentors, but as independent, adaptive, reflective practitioners.

In this sense Schon's notion of reflective practice goes beyond Kennedy's notion of critical analysis noted above. Schon's reflective practice involves both issues of framing and issues of actually dealing with the problems. Sometimes "recipes" apply, but more often for Schon the situation is one of framing the problem, trying out solutions, reframing the problem, trying new solutions, and so on until a reasonable adaptation is reached. In short, reflective practice is a process of knowing rather than a structure of knowledge. The process is one of actually performing in the company and with the help of other professionals.

However, Schon's process of reflective practice seems to many to reduce to trial and error or to remain mysterious. What is it about what good teachers know and do that results in a judgment to choose this book or that, this example or that? How do they decide whether or not to review the unit on fractions one more day or to press on?

Kennedy raises another objection to the concept of reflective practice. Essentially her concern arises over the fact that if we constantly consider both means and ends, ways of understanding the problem and ways of addressing it, we will have no standpoint from which to criticize any particular decision made by a "reflective practitioner." In Kennedy's view the notions of "accountability" or "best practice" seem to be lost. It would seem that "anything goes" and Kennedy quite rightly objects that such a consequence would be unacceptable, both for political and epistemological reasons. The point is that if we allow reflective practitioners constantly to adjust the means and ends they use for addressing a problem, we will be unable to tell whether they have chosen the "right" or "best" or even "better" way of dealing with the problem.

Good judgment

It seems that teachers who are wise in the processes of actually using knowledge have what we call "good judgment." The sense of the term "judgment" which is of interest is the process of deciding what one ought to do or believe, often in the face of uncertainty and changing circumstances. But is judgment, then, a list of propositions? Of judgments written down somewhere? In its primary sense, I think not. There are people who exercise good judgment, but who cannot often write down, or even articulate, any set of propositions which constitutes that judgment. There are other people who do try to write down good judgments. They are the authors of the innumerable "how-to" books which, as we know, at best take us only a very little way toward developing our own good judgment. Converting strategic knowing or judgment into propositionally formulated rules may not be of much help. "Never smile until Christmas, usually." It is the knowledge of when to follow and when to break the rule that constitutes good judgment. As I have been arguing, it is impossible in principle to specify the rule with such completeness that we could ever make the rule explicit. The reason is that judgment is a process, and not basically a proposition.'

The model of judging which I wish to propose has five main interrelated parts (see, for example, Powers, 1973; Sternberg, 1985; Gardner, 1983; and Brown, *et al.*, 1989). People with good judgment have: first, a clear notion of the larger end or ends which they are pursuing; second, the ability actually to perceive situations in terms of the ends which structure those situations; third, the ability to monitor the extent to which their action succeeds or fails in bringing the situation closer to their desired end; fourth, the ability to modify their action so that the situation as experienced gets closer and closer to the desired end; and fifth, the ability to modify jointly the means and ends they are employing in light of the larger social and human purposes they are pursuing.

In more concrete terms this means that teachers who have good judgment usually have a pretty good idea of what they want to accomplish. This notion of ends includes everything from what they may wish students to learn about reading to a view of what kinds of people they want their students to become. Seldom can teachers who view their roles as narrowly instrumental adapt to changing circumstances. The teacher who is confused about ends will be unable to tell whether or not progress is being made.

Teachers with good judgment also have a good knowledge of cases and exemplary practices. They can structure their experience in terms appropriate to the situation. They know, for example, what constitutes misbehavior and what is simply youthful exuberance. Teachers who lack this element of judgment often misinterpret the situation, and, therefore, act inappropriately. How one sees a situation already brings with it a significant tendency towards acting in certain ways.

Monitoring the on-going situation is crucial. Suppose a teacher sees an English class as one in which the students should be gaining experience in textual interpretation, and, therefore, structures the classroom discussion around close textual analysis. However, suppose further that the students do not yet even understand the general purpose of the writing they are supposed to be analyzing. It is crucial for the teacher in such a situation to be able to tell whether the attempts at textual analysis are getting anywhere.

This leads to the fourth element in good judgment—the ability to modify one's actions to achieve the goal. Does the teacher have a repertoire of actions from which to choose? If close textual analysis is not working, simulation or role-playing might help the students interpret the material better. Without a variety of instructional strategies and procedures at their disposal, teachers will be limited to narrow, technically-oriented means to achieve their goals. It is not that one could somehow exercise good judgment in knowing that a given activity was not working, but be totally unaware of alternatives. Judgment means being able to utilize a variety of alternatives. Without the alternatives, there is nothing upon which to exercise judgment.

Finally, the teacher with good judgment can modify jointly the means and ends being used in order to achieve larger social purposes. It may be that what the teacher basically wants to do is to bring the students to a fuller understanding of literature as a liberating force in their lives. In this larger picture, textual interpretation is only one instance of using literature as a liberating force. It may be that with these students, at their stage of development, textual interpretation is inappropriate as a goal and a very different set of means and ends, perhaps relating the story to events in the students' lives, would more adequately serve to liberate them. Note that without this broader adaptability, teachers are likely to blame the students for their inadequacies rather than looking at the larger picture and perhaps realizing that the teachers' own goals and ways of teaching could be modified.

The sense of judgment which I am explicating is not a series of propositions, but a process leading to action. It is not procedural or technical in the sense of following explicit rules laid down by someone else. Rather it is value-laden, and in some cases it is the values themselves which are changed or the emphasis given to competing values altered. But this raises a familiar problem. If we can and sometimes should change the values we are pursuing, then how can we avoid the charge that anything goes? How can we tell good judgment from bad?

The answer is that judgment depends on an evolving tradition to give it point and purpose, and that communal tradition, whether it be conservative or radical, provides the source of the values in terms of which we judge a particular adaptation or alteration of means and ends as good or bad. Furthermore, activity within the tradition is constantly being monitored and adjustments to the tradition made in response to the monitoring.

The fundamental point is that we not only judge the success of teaching or any human activity in terms of how well a given set of means leads to predetermined goals, we also judge those goals by how well the means they call into play allow us to deal with all of the myriad conditions in our human situation. Within the human condition a variety of social arrangements—e.g., law, medicine, science, schooling—have evolved which provide the best means we have been able to create thus far to solve some of our basic human problems. These social arrangements and traditions are themselves subject to judgments, not only of how well and efficiently they operate, but also of how well those particular social arrangements contribute to social well-being. In short, the question is how adaptive are our individual goals and actions and our social organizations, taken together, in allowing us to adapt to the actual ecology in which we human beings find ourselves. The grounds of good judgment are whether or not it leads us to a reflective equilibrium of thought and action (Petrie, 1981, pp. 140-141, 148-150, 180-185, and 213-214).

A good way of testing the account that I have given is to see if there is anything like a notion of "bad" judgment? If judgment requires weighing not only means to predetermined ends, but also, sometimes, those ends themselves, what would constitute "bad" judgment? The answer lies in remembering the essentially social nature of judgment. Judgment is not simply the arbitrary decision of an individual. Rather, there is an unavoidably social nature to the larger norms and standards which inform the reflective practice and the education of reflective professionals.

There are traditions and histories of practice which have evolved over time in ways that have captured at least a modicum of adaptability to the human condition. If they had not, they would have died out. This is, of course, not to say that these traditions have necessarily reached any ultimate truth, but rather to say that they are not simply matters of opinion. They have more or less stood the test of time, and, as such, they are worthy of our respect, however skeptical we, on specific occasions, may be.

"Bad" professional judgment thus turns out to be judgment which does not conform well to the socially and historically developed norms, maxims, practices, and reflections on practice of the profession. One can criticize those norms and practices, but the right to do so must be earned by first learning how to act in accordance with them. Then the criticism will be that the norms and practices, taken as a whole, simply do not really allow us to deal as effectively as they might with that segment of our human ecology with which the profession is supposed to deal.

How does judgment work in specific educational situations? A few examples will help. Research in general pedagogy has confirmed the importance of "direct instruction." The more you actually teach, the more children learn. At the same time, research has also confirmed the usefulness of cooperative learning strategies where children learn from working with each other rather than from being "taught" in any traditional sense. Should a teacher use direct instruction or cooperative learning? The point is that one cannot write down a recipe for when a teacher with good judgment should do one and when the other. Building in even the half dozen or so most common differences in context which ordinary teachers face all the time would make the "recipe" for when to use one and when to use the other impossibly long. Issues of content, age of students, ability level of students, ability of teacher, place in the lesson, place in the curriculum, desire to set the lesson in the context of current events, racial or social background of the class, what the teacher or students had for breakfast, or did not have for breakfast, that morning—all of these and hundreds more play a role in determining whether the teacher with good judgment will use one or the other of the strategies.

Under the analysis I have given of judging, the teacher is constantly monitoring the situation. He or she must know all about the two instructional strategies as well as about the importance of contextual factors. Then, as progress toward learning is perceived to be occurring, the teacher will go ahead with what seems to be working. As the teacher sees the class floundering with a project because of an apparent lack of understanding of a basic concept, the teacher may abandon the cooperative learning strategy and turn to direct instruction to provide a review of the basic concepts underlying the cooperative learning project. Following Shulman, one could expand the example to include judgments concerning particular subjects and their place in the curriculum. Thus, a teacher might have to decide if it is more important within a limited time frame to have students memorize the *Bill of Rights* as fundamental to further learning in American History or if it would be more important for them to understand why the *Bill of Rights* needed to be written.

One can also easily imagine the larger social context influencing judgments made during particular moments of classroom instruction. If the teacher more or less accepts the tradition that the content must be "covered," then in-depth, time-consuming discussion will probably not be attempted very often. However, if the teacher is more interested in promoting critical thought, coverage may be sacrificed. One is reminded of Ted Sizer's (1984) slogan that "less is more." Sizer is precisely committed to cooperatively redesigning the curriculum and structure of his Coalition of Essential Schools to cover less, but with more depth and understanding. The point is not that such schemes are simply changes in the goals schools are meant to serve, which then require a new selection from among the independently validated means we have available for pursuing different ends. Rather, the point is that such basic new conceptions structure our experience in very different ways and will define differently the questions and answers we will pose and seek in our schools and classrooms.

Principles of action in the form of linguistic propositions dealing with all of these situations can at best be provided only in those cases in which the situations are extremely routine, and routine in two distinct ways. First, assuming that the basic values of the tradition are not at issue, the situation must be extremely routine in terms of the instructional context. Introducing a new form of the lever in physics after the basic concept of the lever has been covered may be such a routine situation. Second, one must assume that the basic values of the tradition are not themselves at issue. The same situation which might call for one kind of routine practice assuming that "coverage" is a good thing might well call for a very different kind of routine practice assuming that coverage is not as important as, say, depth.

How does one go about developing good judgment? A full answer to this question is beyond the scope of this paper. Indeed, a full answer would involve laying out a fairly complete prescription for the reform of teacher education. However, a few suggestions derived from the model of judging I have proposed might be in order.

First, it is clear that the development of good judgment will depend upon a knowledge of the ends of schooling and the general place of teaching in society. In short, a good liberal education is necessary. However, the liberal education must be supplemented with work in the foundations of education, for it is here more than anywhere that questions of schooling and society arise. Second, in order to relate thought and action, a far more intensive use of case studies and paradigmatic examples must be made in teacher preparation. Again, foundations will be critical in providing alternative ways of conceiving of the paradigm cases. Third, experience clearly develops good judgment, but only experience in which a wide range of ways of interpreting the situation can be brought to bear in a reflective way. We do learn from experience, but sometimes we learn the wrong things. It is the collective and shared experience of the whole teaching profession from which we should be learning, rather than the individual, often idiosyncratic, experiences of teachers who, because of their isolation, cannot test their interpretations against the collective wisdom of the profession. Although such reflection on experience occurs in a number of places in the typical teacher education program, the foundations typically pay special attention to this aspect.

Teachers must also be equipped with and experienced in a wide range of teaching procedures and techniques in their fields, as well as in the characteristics of diverse learners, so that they will have something upon which to exercise their judgment. Even here, foundational studies, especially history, can provide an antidote to the idea that procedures and techniques are simply a bag of tricks to be acquired and "applied." Finally, as the foundations have always held, teachers must see themselves and be seen as part of the great human and social enterprise of living well. Only if teachers are full participants in the best that is thought and done can they be expected to bring future generations to the level of contributing to the creation of a better life.

It is the great failing of much educational policy-making not to appreciate the morally, socially, and situationally bound context of education and to assume that if only we could analyze teaching in sufficient detail, we could provide the appropriate methods for teaching reading or mathematics or science, that is, specific instructions to be followed by teachers to guarantee learning. In any human activity as complex and morally and socially significant as teaching, that approach simply does not work. In any real-life ecology, there are almost always alternative ways of viewing and understanding situations which bring with them alternative ways of adapting to those situations. Judging wisely involves selecting from among the variety of alternative ways of understanding and dealing with the human and social situations in which we find ourselves.

Accountability

The accountability appropriate to the idea of a professional exercising good judgment is not an accountability of following or not following specific recipes, nor is it even an accountability of student outcomes where the nature of and standards for those outcomes are determined in advance. Indeed, to suppose otherwise was the great failing of process-product research. Such research took the products we wanted as being fixed and they looked for the processes which would lead most effectively and efficiently to those products. Such conceptions simply denied the obvious and important effects of context and variability on the complex situations of teaching and learning. The danger, of course, is that in recognizing the importance of context and variability, one will fall into the unacceptable relativistic position that anything goes.

The conception of accountability which emerges from considering judgment as the relation of research and practice, thought and action, steers the appropriate middle course between predetermined outcomes and anything goes. Furthermore, it is quite congruent with arguments recently being advanced in favor of the professionalization of teaching. Proponents of such professionalization quite correctly point out that the accountability of a profession is lodged not in some independent standard, externally imposed, but in how the profession manages itself within the larger society. This management includes rigorous standards for entry and careful and lengthy preparation, so that those who are accepted have earned the right to criticize the profession and help its further evolution. It also includes the necessity of justifying to society why it should grant the profession such relative autonomy. The challenge for the teaching profession is to convince society that it will *be* better able to pursue its reasonable goals through granting teaching professional status.

It is not simply a political matter as to whether teacher unions have the power to force a professional conception of teaching on society. It rather has to do with the extent to which the norms, maxims, practices, and reflections on practice within teaching have come to define good practice. It is not merely a "knowledge base" which would justify teaching as a profession—at least if that knowledge is understood as a list of propositions. Rather it is a set of informed practices and activities which society can be brought reasonably to believe would further its own goals. In this way, the argument over whether teaching should be professionalized depends heavily on whether we can accept a notion of the relation of theory and practice which itself transcends the non-professional notion of "applying" theory to practice.

Summary

In summary, there are three main implications for teacher knowledge and practice of substituting a notion of professional judgment for that of applying research to practice. First, the notion of applying knowledge to practice all too easily suggests a technical rationality conception of teaching—a view which has come increasingly to be seen as inadequate for thinking about what teachers ought to do. Second, the notion of judgment as the link between research and practice is much more compatible with the conception of teacher as reflective practitioner. In particular, it explains how teachers adapt their knowledge and action to constantly changing situations, but within reasonable parameters set by their profession. Finally, focusing on judgment allows us to begin thinking about professional accountability in much more appropriate ways than checklists of observable teacher behaviors or student scores on standardized tests. It provides a basis for an accountability fully in keeping both with the social responsibility of teachers and with their relative autonomy as professionals.

Notes

1. I have long argued that perceptual learning is a much neglected feature of teaching and learning situations. Indeed, I believe that much of the talk of sensitizing teachers, say, to racial stereotypes, can most fruitfully be addressed as a problem of perceptual learning.
2. See Powers, 1973, for a detailed description of the kind of revolutionary theory of action which gives promise of being able to account for this fact.
3. Shulman's repeated insistence on the value of reflection on practice does not, I believe, count against this point. Reflection on practice is not the same as reflection in order to practice. It is the latter to which I am objecting. Reflection on practice is surely one way in which we improve the whole tradition of the practice.

References

Brown, J., Collins, D., and Duguid, P. (1989). Situated cognition and the culture of learning. *Educational Researcher* 18 (1), 32-42.

Gardner, H. (1983). *Frames of Mind.* Cambridge: Cambridge University Press.

Kennedy, M. (1987). Inexact sciences: professional education and the development of expertise. Issue Paper 87-2, National Center for Research on Teacher Education. Michigan State University. 60 pp.

Petrie, H. (1976). Do you see what I see? The Epistemology of Interdisciplinary Inquiry. *Educational Researcher* 5 (2), 9-15.

Petrie, H. (1981). *The dilemma of enquiry and learning.* Chicago: University of Chicago.[1]

Powers, W. (1973). *Behavior: The control of perception.* Chicago: Aldine.

Schon, D. (1983). *The reflective practitioner.* New York: Basic Books.

Schon, D. (1987d). *Educating the reflective practitioner.* San Francisco: Jossey-Bass.

Schwab, J. (1978). *Science, curriculum, and liberal education: selected essays.* (I. Westbury and N. Wilkof, editors.). Chicago: University of Chicago.

Shulman, L. (1986). Those who understand: knowledge growth in teaching. *Educational Researcher* 15 (2), 4-14.

Shulman, L. (1987a). Assessment for teaching: an initiative for the profession. *Phi Delta Kappan* 69 (1), 38-44.

Shulman, L. (1987b). Knowledge and teaching: foundations of the new reform. *Harvard Educational Review* 57 (1), 1-22.

Sizer, T. R. (1984). *Horace's compromise: the dilemma of the American high school.* Boston: Houghton Mifflin.

Sternberg, R. (1985). *Beyond IQ: A triarchic theory of human intelligence.* Cambridge: Cambridge University.

---

[1] Revised and expanded (2011) Menlo Park, CA: Living Control Systems Publishing

[1995]
A New Paradigm for Practical Research

Why has nearly a century or more of educational research been of so little help in actually improving teaching and learning? Why is educational research preoccupied with "method" and methodological disputes to a much higher degree than "real" science? With the increased attention during the last several years to inquiry in authentic educational settings, such as professional development schools, what kind of educational research is likely to make a difference? How will we know?

There are doubtless any number of answers to these questions, but in this chapter, I want to explore one possible line of thought that can both account for the paucity of serious research results in teaching and learning as well as suggest why certain kinds of situated and context-dependent inquiry in professional development schools hold much more promise of actually making a difference.

It is important that I say what I am not going to do. I am not going to consider the vast quantity of philosophical, historical, political, and sociological analyses of various aspects of education. Much of this work has been of significant importance in increasing our understanding of education and may well escape the critique I will offer. Nor am I going to conduct an exhaustive review of the literature to cover all of the claims and counterclaims and methodological disputes and controversies surrounding educational research.

Rather, I want to concentrate on the core psychological and social psychological research that has long been taken by many to be the key to understanding teaching and learning and how to make them better. I will examine in some depth four representative analyses of research into teaching and learning over the past several decades and consider why these penetrating critiques have not, at least up until now, had much effect on our conduct of educational research into teaching and learning.

First published in: Petrie, H.G. (ed.), Professionalization, partnership, and power: building professional development schools. Albany, NY: SUNY Press.

What I will suggest is that the theories of learning and behavior of the past 100 years or so, from behaviorism to constructivism, have relied on a flawed conception of human nature. However, as Kuhn (1970) clearly showed, until a more reasonable conception comes along, all of the problems in a research paradigm will be "solved" by adding complications to the existing theory or by that most ubiquitous of journal article endings, "more research is needed!" Just as the Ptolemaicists added epicycle upon epicycle to save their earth-centered theory of the universe from constant anomalies, so also do traditional psychologists protect their core beliefs by adding ad hoc assumption after ad hoc assumption. Indeed, the psychologists are even worse. To compensate for an inadequate fundamental conception of human nature, they have utilized the sophisticated mathematical discipline of statistics to explain away what would otherwise be plain to everyone as wholly inadequate accounts of human behavior.

The impotence of critique

Upon the occasion of his receiving the Distinguished Scientist Award of Division 12, Section 3 of the American Psychological Association, Paul Meehl delivered a lecture later published (1978) with the revealing title, "Theoretical Risks and Tabular Asterisks: Sir Karl, Sir Ronald, and the Slow Progress of Soft Psychology." In this article, Meehl argued forcefully that one of psychology's major mistakes has been in trusting too much in Sir Ronald (Fisher) and his notions of significance testing and too little in Sir Karl (Popper) and his theory of falsification.

The point is well known in philosophy of science. If one is trying to test a scientific hypothesis (H), in conditions (C), using auxiliary apparatus and methods (A), then typically one tries to predict from the conjunction of these situations some observation (O), that one believes one can make. That is, if H and C and A, then O. The problem is that if one actually does observe what is predicted, there is nothing, logically, that can be concluded, for the observation might well have been an accident. To conclude that H is, in fact, true, would be to commit the logical fallacy of affirming the consequent. This is the well-known "paradox of confirmation" (Carnap, 1950).

On the other hand, as Popper (1959, 1962, 1972) has pointed out, if the predicted observation does not occur, that is, if not-O, then a very strong conclusion can be reached via the logical inference of modus tollens.

Either the hypothesis is not true or the conditions did not hold, or the auxiliary apparatus and methods are wrong. In the hard sciences (perhaps this is why they are hard), there is seldom any doubt that the conditions did hold and the apparatus, for example, measuring devices, are based on well-established theories. Consequently, one can usually conclude that the hypothesis is false. From these facts, Popper develops his notion of falsification. The important thing for a science to do is to expose its hypotheses to strong tests that would tend to falsify them. Platt (1973) elaborated this into the notion of "strong inference" which suggests that sciences that are really on the right track seldom, if ever, engage in significance testing. Rather they elaborate alternatives to account for the phenomena and quickly proceed, through exposing them to strong possibilities of falsification, to eliminate them. The ones that withstand this process of falsification have some real claim to validity and are embedded into the well-tested core of the science.

Not so with the soft sciences of psychology. Instead, as Meehl (1978) so aptly pointed out, hypotheses in psychology "suffer the fate that General MacArthur ascribed to old generals—they never die, they just slowly fade away" (p. 807). Meehl listed some 20 intrinsic difficulties in making psychology into a real science. I shall return to one of them—the problem of intentionality—below, but his major target is the poor way of doing science represented by the overwhelming reliance on significance testing instead of falsification. He said:

> I believe that the almost universal reliance on merely refuting the null hypothesis as the standard method for corroborating substantive theories in the soft areas is a terrible mistake, is basically unsound, poor scientific strategy, and one of the worst things that ever happened in the history of psychology (p. 817).

Essentially, Meehl's argument is that because of the extremely complex nature of human behavior, it will never be possible to make certain that all of the potentially contributing factors to any result are either equal or properly counterbalanced. For example, test scores of students taught by whole language versus phonics may be due to the treatment *or* to one or another of the complexities of the conditions affecting individual students. Consequently, with any reasonable set of measures the null hypothesis will *always* be falsified, but we will never know whether or not it is because the substantive hypothesis is actually true or because of one or more of the complexities in the antecedent conditions. Consequently, the *inevitable* result in psychological theorizing with significance testing will be that we will have mixed results.

Why then, in the face of this powerful methodological critique, do we still find journal articles and graduate methodology courses religiously using significance testing? Even worse, why do we eyeball the tables of results and count up the places in which there is a significant difference and those in which there is not and somehow conclude something substantive about what has occurred? Meehl's answer is that the hard scientist

> has a sufficiently powerful invisible hand theory that enables him to generate an expected curve for his experimental results. He plots the observed points, looks at the agreement, and comments that 'the results are in reasonably good accord with theory.' Moral: *It is always more valuable to show approximate agreement of observations with a theoretically predicted numerical point value, rank order, or function form, than it is to compute a 'precise probability' that something merely differs from something else* (1978, p. 825).

What does Meehl mean by an "invisible hand theory?" I suggest that he is referring to an underlying model, which, if it operates as hypothesized, would yield the predictions. This would be like the kinetic theory of gases, or electromagnetism, or chemical bonding. These models in the hard sciences allow us to make precise predictions of observed features of temperatures, meter readings, and chemical reactions. We have historically had nothing remotely resembling such a model in the soft science of psychology, and the bare bones empiricism of speaking of stimuli and responses or independent and dependent variables succeeds at best in redescribing the phenomena rather than providing an explanatory theory. Consequently, despite the power of Meehl's critique of significance testing, in the absence of any notion of a powerful generative model, psychologists continue to rely on statistical significance testing.

Three years before Meehl's 1978 article, Lee Cronbach (1975) published, "Beyond the Two Disciplines of Scientific Psychology" based on his Distinguished Scientific Contribution Award from the American Psychological Association. In critiquing the paucity of results from aptitude-treatment interaction (ATI) research that he himself had strongly advocated nearly twenty years earlier, Cronbach asked, "Should social science aspire to reduce behavior to laws?" (p. 116).

Cronbach, too, accepts the basic formulation of the problem facing psychology to be that of predicting observed behavior from the conjunction of hypothesized laws, initial conditions, and experimental apparatus. If H and C and A, then O. The aptitude-treatment interaction line of research is essentially an attempt to specify the various conditions, C, under which different observations might be predicted. For example, if one's instructional hypothesis, H, is that students learn better if they are challenged by the instructor, ATI research suggests that this tends to be true under conditions, C_1 where the student has the personality type to seek challenges and accept responsibility, but not so under conditions, C_2, where the student is more defensive.

But, as Cronbach (1975), pointed out, the potential number of conditions that might need to be considered is limitless.

> If Aptitude × Treatment × Sex interact, for example, then the Aptitude × Treatment effect does not tell the story. Once we attend to interactions, we enter a hall of mirrors that extends to infinity. However far we carry our analysis, to third order or fifth order or any other, untested interactions of a still higher order can be envisioned (p. 119).

The complexity of seemingly limitless potential interactions was not all that troubled Cronbach. He also suggested that generalizations decay over time and psychological generalizations decay more rapidly than do generalizations in the physical sciences. This is due to a number of factors, chief among which is the changeable nature of the social and psychological world, that render any hypotheses we might suggest valid for only a very short period of time.

> Our troubles do not arise because human events are in principle unlawful; man and his creations are part of the natural world. The trouble, as I see it, is that we cannot store up generalizations and constructs for ultimate assembly into a network. . . . If the effect of a treatment changes over a few decades, that inconsistency is an effect, a Treatment × Decade interaction that must itself be regulated by whatever laws there be (p. 123).

There are regularities of the "if H and C and A, then O" variety in psychology according to Cronbach, but they change so rapidly that we can never assemble them into a theory. So what did he suggest?

> Instead of making generalization the ruling consideration in our research, I suggest that we reverse our priorities. An observer collecting data in one particular situation is in a position to appraise a practice or proposition in that setting, observing effects in context. In trying to describe and account for what happened, he will give attention to whatever variables were controlled, but he will give equally careful attention to uncontrolled conditions, to personal characteristics, and to events that occurred during treatment and measurement. As he goes from situation to situation, his first task is to describe and interpret the effect anew in each locale, perhaps taking into account factors unique to that locale or series of events. . . . (pp. 124-125)

So *context* and the interpretation of events in context become important to the psychologist. We must pay particular attention to the variability of human action in even one actor, let alone across different individuals and how (presumably) the *same* effect can occur anew in differing and unique locales. Yet, despite the trenchant critique of seeking if-then lawlike generalizations, Cronbach continues to hold to the position that there are such laws; we just cannot discover them quickly enough. Consequently, those who would take an anthropological approach to interpreting human behavior are still not quite fully scientific. They are just doing the best that they can.

Going back even farther, I investigated another reason for the paucity of educational results emanating from psychology in my paper, "Why Has Learning Theory Failed to Teach Us How to Learn," (Petrie, 1968). The major burden of the argument was that classical stimulus-response theory, along with most then-extant variants, was simply inadequate to account for human intentionality. Furthermore, educational practice is shot through and through with presumptions that human behavior is fundamentally intentional in character. That is, teachers, students, administrators, parents, policy makers, indeed, all of us, do things on purpose, in order to pursue certain goals, not because of the operation of some "if-then law," no matter how complex or short its half-life might be.

Thus, learning theory had failed to teach us how to learn, I argued, because it was, in an essential way, talking about something quite different from what educators were talking about. The solution I proposed at that time was either to recast education into stimulus-response terms (something I felt had little chance of success) or find a way in which our psychological theorizing could take account of human intentionality. In the language of the time, the problem was to show how reasons could be conceived of as causes, not simply "if-then" causes, but rather "in order that" causes.

I would now slightly rephrase the point that learning theorists and educators are essentially talking about different things. The fundamental interest of the educator is in the individual student. What can I do to help Suzy or Johnny learn about diverse cultures? How can I do this when I know that Suzy comes from a bigoted family and Johnny is the son of a biracial couple? The fundamental interest of the learning theorist is in the laws of learning as they apply to all students. In principle, these interests of the educator and the learning theorist could overlap considerably. It is logically possible that the laws of learning apply to all individuals, just as the laws of mechanics apply to all point masses.

However, it is clear by now that this logical possibility has not been realized in practice. We have no degree of real assurance that any laws of learning apply to individuals. Any given person may or may not react as predicted. This could be due to a variety of reasons. It might be that human behavior is simply so complex that, although the laws of learning do apply, the complexities of the situation preclude our being able to use them with any reliability. To continue the mechanics analogy, it would be like trying to predict when a given leaf from a tree would fall in the autumn and what path it would take in reaching the ground. All of those phenomena are clearly governed by the laws of mechanics, but the situation is too complex to allow for any meaningful prediction.

My suspicion is that most learning theorists implicitly assume this tack. This approach maintains the hegemony of traditional psychology and learning theory in educational research, while explaining away the lack of any more useful guidance than has been forthcoming. Furthermore, there are just enough semiuseful statistical generalizations to allow teachers to "apply" learning theory in their classrooms and reach some kind of success with at least a portion of their students. They are, however, completely befuddled by why they are unsuccessful with the rest. The learning theorists, on the other hand, as we have seen with Cronbach, paint a picture of very complex laws, highly dependent on individual differences, which might, nonetheless, ultimately yield to more sophisticated investigations.

An alternative explanation of the difference between learning theorists and educators is that educators, in their emphasis on the individual, intuitively know that not only do different individuals behave differently, but that the same individual will vary his or her behavior in varying circumstances in order to reach consistent ends. In short, educators recognize what conventional learning theorists typically do not, that individual behavior is fundamentally purposive and intentional. The question then becomes not one of increasing the complexity of traditional learning theory accounts but of giving a persuasive account of how intentional action on the part of students and teachers is possible.

To put the problem in terms of the preceding discussion, we need to come up with a generative model of human behavior, an invisible hand theory in Meehl's terms, that can account for the seemingly indefinite number of ways in which human beings can pursue their goals in the face of the kind of constantly changing circumstances noted by Cronbach. Furthermore, if at all possible, the model should not be that of infinitely complex "if-then laws" that would cover all of the possible interactions.

In many respects, the current so-called constructivist theories of behavior (e.g., Brown, Collins, & Duguid, 1989) have taken seriously the challenge to account for human intentionality. We construct meaning out of our individual experience on the basis of our wants, desires, needs, and the limitations imposed by the physical and social environments.

However, it is not clear that the recent spate of constructivist theories has yet broken free of the linear causation implied by the "if-then" form of causal laws. The constructivists still tend to phrase their speculations in the form of, "If I *intend* to get students to learn to read and *I believe* in the efficacy of phonics instruction, then I will drill students on the various sound combinations." This is, of course, tremendously oversimplified, but it illustrates the kind of linear causation from intention and belief to behavior.

But what has not happened, for the most part, is the creation of invisible hand or generative models of behavior that would allow the specific, "point predictions" spoken of by Meehl. There is one promising exception that I will note later, but in general we are still at a very primitive level in our psychological theorizing. In fact, most of the "theorizing" is probably not much more than a kind of "explanation by redescription." That is, our theories are seldom much more than describing consistent phenomena with pretentious language and claiming that we have constructed a theory. "Why isn't Johnny paying attention?" "Oh, he has attention deficit disorder." "What is that?" "Attention Deficit Disorder is the tendency not to pay attention."

Faced with this kind of impasse in traditional approaches to psychological theorizing, it is no wonder that some educational researchers simply throw up their hands at any possibility of finding an underlying theory of human behavior and turn to exploring what they call alternative forms of understanding. In his 1993 AERA Presidential Address, "Forms of Understanding and the Future of Educational Research," Elliot Eisner explored just these notions. He built upon his deep experience with the arts to argue for multiple conceptions of ways of knowing related to multiple forms of representation that different people bring to experience.

Eisner (1993) paid particular attention to how we learn to experience the world. He argued persuasively that perception or experience is not some neutral given upon which we impose interpretations; rather the very substance of what we experience is the result of an interaction of mind and sense. "I came to believe that humans do not simply have experience; they have a hand in its creation, and the quality of their creation depends upon the ways they employ their minds" (p. 5).

It is important to note, as Eisner (1993) said,

> In talking about experience and its relationship to the forms of representation that we employ, I am not talking about poetry and pictures, literature and dance, mathematics and literal statement simply as alternative *means* for displaying what we know. I am talking about the forms of understanding, the *unique* forms of understanding that poetry and pictures, literature and dance, mathematics and literal language make possible (Emphasis added, p. 8).

In short, Eisner is suggesting that traditional psychology, with its impoverished conceptual schemes, simply cannot account for the multiple experiences that we have. Consequently, we need to simply cordon off these multiple modes of experience and grant them their own autonomy as ways of knowing.

To this end, Eisner suggests a number of changes that might take place in educational research. We would probably see an expansion of research methods, for example, an increasing use of narrative and poetic forms of research. Furthermore, such an expansion would likely have an effect on the ways in which we teach various subjects. We would be more likely, for example, in teaching history to make use of music, architecture, film, stories, and the like, not only as parts of traditional lectures but as unique ways in which they can shed light on the history in question. Student demonstrations of competence would also differ. We might see some preparing a video, others writing a poem, still others engaging in some action project, and some continuing to demonstrate their competence through multiple choice exams and traditional dissertations.

Carrying the speculation even further, Eisner asks what might the presentation of educational research look like? Would novels count? A multimedia presentation? An MTV video? How might all of this be judged?

Yet, there are difficulties that this kind of "multiple forms of understanding" approach raises. How, for example, is it even possible for us to understand what Eisner is proposing, unless we have some overarching form of understanding in terms of which we can see and appreciate the possibility that poems might complement or supplement what we can learn from descriptions and numbers? In other words, if these different forms of understanding are *completely* different, how will we ever be able to integrate the understandings that each provides into some sort of human whole? Will we each be segregated into a literal self, a numerical self, a poetic self, an artistic self? As Eisner (1993) himself asked, "Can we translate what is specific and unique to forms other than those in which such understanding is revealed?" (p. 10)

So, despite the allure of the concept of multiple forms of understanding, we seem to be driven back toward some general notion of human understanding and behavior that could account for our ability to engage in and understand these different forms. We need, apparently, to be able to see these multiple activities and understandings in a unified way *both* as something human beings do *and* as diverse activities within that understanding. Once again the critique of traditional psychological theorizing can be seen as important but not quite powerful enough to turn the tide.

A possible synthesis

What, then, are the major challenges facing psychological educational research? From Meehl we get a powerful critique of traditional significance testing as a way of deciding among hypotheses. He also suggests that we need a generative model (invisible hand theory) that can underlie and account for the surface predictions we do make. From Cronbach we see the indefinite complexity that appears to attend the attempts to get traditional if-then laws to account for individual human behavior. He suggests that we must always take context into account. I have argued that human purposefulness and intentionality must be the central feature for which any theory of behavior must account. Eisner reminds us of the incredible variety of ways of dealing with our world we human beings employ and of just how central is the function of how we perceive or experience that world.

Interestingly, over the past 20 years or so, there has begun to emerge a small, still highly controversial, body of work that promises to meet all of the challenges to psychological educational research enumerated above. This conception of human behavior was given its most powerful formulation by W. T. Powers (1973) in his book, *Behavior: the Control of Perception*. Most recently in education, it has been the subject of a spirited debate in the pages of *Educational Researcher* (Cziko, 1992a, 1992b; Amundson, Serlin, & Lehrer, 1992). There are also a number of other researchers from a variety of disciplines contributing to this body of research in psychology (Powers, 1989; Robertson and Powers, 1990), experimental psychology (Bourbon, 1990; Hershberger, 1988; Marken, 1986, 1989, 1990, 1992), clinical psychology (Ford, 1993, 1994; Goldstein, 1990), education (Bohannon, Powers, & Schoepfle, 1974; Petrie, 1974, 1979, 1981), management (Forssell, 1993), sociology (McClelland, 1994; McPhail, 1991; McPhail, Powers, & Tucker, 1992), ethology (Plooij & van deRijt-Plooij, 1990), law (Gibbons, 1990), and economics (Williams, 1989, 1990).

This new conception of human nature is called perceptual control theory and, as the title of Powers' book implies, it fundamentally turns our conceptions of human nature on their heads. Instead of viewing behavior as the outcome of stimuli or perceptions (as modified by cognition, emotions, or planning), perceptual control theory views behavior as the means by which a perceived state of affairs is brought to and maintained at a (frequently varying) reference or goal state. Perceptual control theory escapes the problem of modeling behavior as planned and computed output, an approach that requires levels of precise calculation that are unrealistic in a physical system and impossible in a real environment that is changing from one moment to the next. Instead, perceptual control theory provides a physically plausible explanation both for the consistency of outcomes of human action and the variability of means utilized to achieve those outcomes in a constantly changing environment.

Perceptual control theory makes use of the "circular causation" found in engineering control and servo-mechanism theory. Thermostats and cruise control systems are everyday examples of mechanical control systems that keep the perception of temperature or speed near the reference levels set for them. Many people, when hearing of these engineering control systems as examples, are immediately put off by perceptual control theory, thinking that it must be a highly mechanistic theory. Nothing could be further from the truth. Engineering control systems arose precisely from the problem of wanting to create mechanical systems that behaved like human beings as we might go about the tasks of governing temperature, maintaining speed, tracking targets and so on.

Paradoxical as it may seem, traditional psychology with its emphasis on if-then, stimulus-response, input-output, independent-dependent variable kinds of laws and its efforts to model itself on physics adopted the truly mechanistic view of behavior. On the other hand, engineers, unencumbered by worries about psychology and interested only in obtaining performances from mechanical systems analogous to what real people can do, were able to create a theory that is much more amenable to modeling actual human behavior than those created by the psychologists.

It is not possible in this short chapter to give a complete introduction to perceptual control theory. Cziko (1992a) gives a brief introduction and the classic is still Powers' (1973) wide-ranging and very readable presentation. What I will do here is provide a very brief sketch of how perceptual control theory begins to answer the major challenges to traditional psychology outlined previously.

Meehl's critique of significance testing and call for appropriate "invisible hand" models are met head-on. In the areas in which they have been tested, generative models based on perceptual control theory have been developed that correlate with the actual point by point behavior of individual subjects at values between .97 and .99 (e.g., Bourbon, 1990; Marken, 1986, 1989, 1992). These are, furthermore, real "invisible hand" models, in that, once built, they predict entirely novel behavior in situations not before encountered.

This capacity of the theory relates to Cronbach's concerns with the seemingly indefinite number of variables that might enter into any of the more traditional laws of learning. Consider the mechanical cruise control system. There are an indefinite number of factors that might keep a car from maintaining a certain speed—headwinds, crosswinds, hills, curves, poor quality gasoline—the list is endless. If we tried to build a mechanical system that would be able to determine when any of these features might interfere with the desired speed (which desired speed itself might change from time to time during a trip), and also include the capacity for calculating just how much gas to deliver to the engine to overcome any of these disturbances, we might well conclude that the "individual differences" in the mechanical case were every bit as daunting as Cronbach concluded they were in the human case.

But that is not what the engineers did. Rather, they built a mechanism that sensed the speed of the car, compared that speed to the desired one, and if it was too slow, fed more gas to the engine and if too fast, decreased the gas. The cruise control *does not know and does not care* what causes the speed to depart from its desired level, it just compensates for it when it does. In short, it controls the perception of the speed of the car, keeping it very close to the desired level, and it does this in a "circular causation" kind of way in which the output affects the input at the same time as the input is being compared to the reference speed and the difference between the two is actuating the output.

Note, too, that this is just how the human driver without cruise control behaves. We do not check headwinds or hills, especially if they are slight, and compute how much to depress or let up on the accelerator. Rather, we monitor the speedometer and no matter what the cause of a change in the speed we want to maintain, we depress or let up on the accelerator accordingly.

Similarly with learning. The expert teacher (e.g., Berliner, 1989) does not calculate what to do to counteract each disturbance to a child's learning. Rather, the expert senses the difficulties the child is having and in a flowing way adjusts to the situation. Indeed, such an ability to sense the teaching act at this more abstract level is precisely what distinguishes the more expert teacher from the novice who mostly relies on mechanical step-by-step recipes. In short, the expert teacher has a reference level for students learning a particular concept or fact and is constantly comparing the perception of the students' performance with that level and varying outputs to bring the teacher's perception into congruence with the reference level for learning. The expert teacher no more needs to know the detailed "laws of learning" than does the cruise control system need to know the physics of how an incline will slow down the momentum of the car.

Context is, indeed, all important, but a control system does not, in most cases, need to sense the context in order to take it into account. Context is simply another name for the myriad differences in a constantly changing environment. These changes act as disturbances to the perceived variable (e.g., the speed of the car or the learning of the student) that is being controlled. Unless the disturbances are overwhelming, good control systems sense the difference between what they desire to perceive and what they are perceiving and automatically behave in ways tending to counteract the disturbance.

A central function of perceptual control theory is to account for intentionality and purpose. Control systems are precisely organized to allow a consistent end to be reached with varying means in a constantly changing environment. If, on my trip to the office, I find a street blocked off, I find another way, even if I have never gone that way before. I usually do not even need a "detour" sign to tell me what to do. I can simply vary my behavior appropriately in these changed circumstances to achieve my goal. I can also vary my proximal goal of getting to the office in light of the higher order goal of stopping to help an accident victim.

Similarly with the expert teacher. If the books for the students have not arrived, adjustments can be made. In order to achieve the overall goal of understanding the United States constitution, the teacher can throw out the original lesson plan and adapt the discussion to take advantage of newspaper accounts of Russia changing its constitution. If an earthquake requires attention to immediate student fears, the teacher readjusts the lessons accordingly in view of the higher order goal of caring for the students in a crisis.

Furthermore, the conception of behavior as the control of perception gives a transparent account of why, as Eisner reminds us, our perceptions of the world are so important. What human beings *do* in the world is control their perceptions. Coming to understand or be competent in a "way of knowing" is, on this account, coming to be able to recognize and control those kinds of perceptions. The .300 hitter in baseball can *see* the ball better than others. The chess grandmaster *perceives* strength in the middle. The astute social commentator *senses* the breakdown of family life. The "with it" teacher has *eyes* in the back of her head. The artist *observes* the world with more clarity than do the rest of us.

The vast range of ways we humans have of dealing with the world is, indeed, remarkable. From the standpoint of perceptual control theory, however, what is truly remarkable is that this diversity of means in the face of a constantly changing environment is usually for the sake of achieving the same consistent ends. Perceptual control theory shows us how we make the one out of the many, how we find the *unum* in the pluribus.

Finally, perceptual control theory is fundamentally a theory of how individuals behave. In this respect, it is precisely the kind of theory that teachers need. They need to know about individual students who face them in their classrooms, not about what kids learn or do "on the average." The problem with the average is that it often washes out the interesting stuff. Because of both individual differences in children and differences in the means employed by a single child across time to achieve consistent ends, it becomes obvious that it is not a question of either praise or blame for a student's behavior, but rather a question of when to use what for which child.

Professional development schools

If perceptual control theory does supply a superior model for understanding human behavior, then it does so in the laboratory and the classroom as well as in the professional development school. Nevertheless, the professional development school notion seems particularly congenial to practical research conducted from the standpoint of perceptual control theory.

The professional development school is a place in which best practice is to be modeled, learned, and evaluated. It is a profoundly practical site, a place where teachers, students, administrators, other educators, and professors come together to try to figure out what to do when faced with specific goals and challenges in a specific environment. It is a place where "theory is put into practice."

Yet, the insight of perceptual control theory is that there are no overall theoretical laws of human behavior governing what people always do in given circumstances. Such laws would only be possible in an extremely stable universe, which is certainly not the one in which we live. The very concept of putting theory, perhaps generated in an orderly laboratory, into practice in a disorderly world is wholly inappropriate. Indeed, we can see that theory, at least as traditionally conceived, could not possibly work in practice. The only possibly correct theory would have to take into account the fundamental fact that human beings are able to achieve consistent ends in indefinitely varying circumstances. The "laboratory" *must* be the real world because human nature is such as to be able to pursue our goals in the real, constantly changing world.

There is also a very immediate reason why a professional development school would be a natural place for research on teaching and learning. Under perceptual control theory what people learn to do is to control their perceptions, not necessarily to perform certain routinized actions. Thus, the most effective teaching is likely to be providing students with exemplary perceptions of what it is they are trying to learn, not detailed instructions on how to get there (Petrie, 1974, 1986; Petrie & Oshlag, 1993). We should, for the most part, show students what the finished product should look like, rather than give them recipes to follow. It is usually much easier to show aspiring teachers, for example, what good teaching looks like in a real school rather than to describe it in the university classroom.

One way of interpreting the common wisdom regarding quantitative and qualitative research is that individual qualitative research might suggest hypotheses that must then be seriously confirmed quantitatively through large samples. Perceptual control theory turns this common wisdom on its head. At best, quantitative research on large groups of people might suggest general tendencies that may hold true in limited circumstances. But those suggestions would have to be confirmed with individual teachers and students living in an ever-changing world. Traditional psychological research may provide a few hints about how to start real research on teaching and learning in real contexts.

The professional development school, then, is the real laboratory. The aspiring teachers and administrators and counselors and psychologists and social workers need to be provided with examples of the perceptions of learning, classroom order, cooperation, and the like, that they are learning to control. They need to learn how to determine their own goals and those of their co-workers as well as those of students and parents. They need to practice the skills of situational analysis and consensus building so that all can control successfully as much of their perceptual worlds as possible. They need to come to respect others as persons. What this means is recognizing that others, like oneself, are control systems who will try to resist disturbances to the perceptions they are trying to control. It means understanding that the only way actually to control others is through overwhelming physical force and then only until they find a way of evading the force. It means searching for ways of looking at and dealing with the world that can allow for a maximum of mutual satisfaction.

In an extremely important way, human beings are even more predictable than are physical events. Human beings are organized to attain consistent goals *despite* varying circumstances. An automobile mechanically programmed to drive around a given track will be less predictable in its path in the face of significant crosswinds than will the path of that same automobile in the hands of a human driver wanting to drive around the track.

When we know what people want, we know that they will do what they have to do to attain what they want. That is what control systems do. We do not know just *how* they will get to their goal, but we do know that they will likely achieve it, assuming it is within normal ranges of possibility. If I want a drink, we can predict I will get it, even if we cannot predict whether I will use a cup, a glass, my hands, or just stick my head under the spigot.

Of course, we also know that people's wants and desires are complicated and interrelated. Some things that we desire are desired in order to attain higher-order goals. I want to go to the office in order to work in order to do something I enjoy and find worthwhile that is consistent with my view of the kind of person I want to be. Some things that I want at one time, I do not want at another time, or I want less of them. As much as I like chocolate, I do not want only it nor do I want it all the time, probably because I also want to live a reasonably healthy life.

There are, of course, many, many issues to be explored regarding how our wants and desires fit together or fail to fit together. What roles do memory and imagination and hallucination play? How do independent control systems interact with each other in social arrangements? How do we learn or change our control systems when we persistently fail to be able to control our perceptions? How do we find out what other people want so that we can begin to understand how to interact with them? These and a host of other questions suggest themselves.

What is critical for this discussion, however, is that perceptual control theory gives us a very different perspective on the kind of practical knowledge educators need. Traditional psychology with its presumptions of if-then laws of behavior can maintain its hegemony over practical research as long as there is the hope of actually finding such laws. No matter how complicated such laws may be, if human beings really are subject to them, the professional development school and its kin will be seen as the place in which these laws are applied, not discovered. If, on the other hand, we have an alternative conception of human action as controlling perception rather than being controlled by it, as being purposeful and able to attain consistent goals by varying means in a constantly changing environment, then we will recognize that these achievements can only be explained by the operation of a control system. In that event the professional development school will be seen as the very place in which individuals as autonomous control systems learn about each other and how they can coexist and mutually satisfy their needs and wants.

There are no "laws of learning," at least as that phrase is ordinarily understood. There are only the laws governing the way in which human beings are organized and how they can come to be reorganized, and those laws predict exactly the autonomous goal-seeking, yet variable behavior we see. For those of us who would educate educators, professional development schools are where those laws of human organization and reorganization are most fully on display.

References

Amundson, R., Serlin, R. C., & Lehrer, R. (1992). On the threats that do not face educational research. *Educational Researcher, 21(9),* 19-24.

Berliner, D. (1989). Implications of studies and expertise in pedagogy for teacher education and evaluation. *New directions for teacher assessment: Proceedings of the 1988 ETS international conference,* 39-78.

Bohannon, P., Powers, W. T., & Schoepfle, M. (1974). Systems conflict in the learning alliance. In L. J. Stiles (Ed.), *Theories for teaching* (pp. 76-96). New York: Praeger.

Bourbon, W. T. (1990). Invitation to the dance: Explaining the variance when control systems interact. *American Behavioral Scientist, 34(1)* 95-105.

Brown, J., Collins, D., & Duguid, P. (1989). Situated cognition and the culture of learning. *Educational Researcher, 18(1),* 32-42.

Carnap, R. (1950). *Logical foundations of probability.* Chicago: University of Chicago Press.

Cronbach, L. J. (1975). Beyond the two disciplines of scientific psychology. *American Psychologist, 30,* 116-126.

Cziko, G. A. (1992a). Purposeful behavior as the control of perception: Implications for educational research. *Educational Researcher, 21(9),* 25-27.

Cziko, G. A. (1992b). Perceptual control theory: One threat to educational research not (yet?) faced by Amundson, Serlin, and Lehrer. *Educational Researcher, 22(7),* 5-11.

Eisner, E. (1993). Forms of understanding and the future of educational research. *Educational Researcher, 22(7),* 5-11.

Ford, E. E. (1993). *Freedom from stress.* Scottsdale, AZ: Brandt Publishing.

Ford, E. E. (1994). *Discipline for home and school.* Scottsdale, AZ: Brandt Publishing.

Forssell, D. C. (1993). Perceptual control: A new management insight. *Engineering Management Journal, 5(4),* 1-7.

Gibbons, H. (1990). *The death of Jeffrey Stapleton: Exploring the way lawyers think.* Concord, NH: Franklin Pierce Law Center.

Goldstein, D. M. (1990). Clinical applications of control theory. *American Behavioral Scientist, 34(1),* 110-116.

Hershberger, W. A. (Ed.). (1988). *Volitional action: Conation and control.* Amsterdam: Elsevier.

Kuhn, T. S. (1970). *The structure of scientific revolutions* (2nd ed.). Chicago: University of Chicago Press.

Marken, R. S. (1986). Perceptual organization of behavior: A hierarchical control model of coordinated action. *Journal of Experimental Psychology: Human Perception and Performance, 12,* 267-276.

Marken, R. S. (1989). Behavior in the first degree. In W. A. Hershberger (Ed.), *Volitional action: Conation and control.* Amsterdam: Elsevier.

Marken, R. S. (1992). *Mindreadings: Experimental studies of purpose.* Gravel Switch, KY: CSG Books.

McClelland, K. (Winter, 1994). Perceptual control and social power. *Sociological Perspectives, 37,* 461-496.

McPhail, C. (1991). *The myth of the madding crowd.* New York: Aldine DeGruyter.

McPhail, C., Powers, W. T., & Tucker, C. W. (1992). Simulated individual and collective action in temporary gatherings. *Social Science Computer Review, 10(1),* 1-28.

Meehl, P. (1978). Theoretical risks and tabular asterisks: Sir Karl, Sir Ronald, and the slow progress of soft psychology. *Journal of Consulting and Clinical Psychology, 46(4),* 806-834.

Petrie, H. G. (1968). Why has learning theory failed to teach us how to learn. In G. Newsome (Ed.), *Philosophy of Education 1968* (pp. 163-170). Lawrence: University of Kansas Press.

Petrie, H. G. (1974). Action, perception, and education. *Educational Theory, 24,* 33-45.

Petrie, H. G. (1979). Against 'objective' tests: A note on the epistemology underlying current testing dogma. In M. N. Ozer (Ed.), *A cybernetic approach to the assessment of children: Toward a more humane use of human beings* (pp. 117-150). Boulder, CO: Westview Press.

Petrie, H. G. (1981). *The dilemma of enquiry and learning.* Chicago: University of Chicago Press.[1]

---

1 Revised and expanded (2011) Menlo Park, CA: Living Control Systems Publishing

Petrie, H. G. (1986). Testing for critical thinking. Presidential address, *Proceedings of the Philosophy of Education Society,* 3-20.

Petrie, H. G., & Oshlag, R. S. (1993). Metaphor and learning. In A. Ortony (Ed.), *Metaphor and thought* (2nd ed.). Cambridge: Cambridge University Press, 579-609.

Platt, J. (1973). Strong inference. In H. S. Brody, R. H. Ennis, & L. I. Krimerman, (Eds.), *Philosophy of Educational Research* (pp. 203-217). New York: Wiley.

Plooij, F. X, van deRijt-Plooij (1990). Developmental transitions as successive reorganizations of a control hierarchy. *American Behavioral Scientist,* 34(1), 67-80.

Popper, K. (1959). *The logic of scientific discovery.* New York: Basic Books.

Popper, K. (1962). *Conjectures and refutations: The growth of scientific knowledge.* London: Routledge and Kegan Paul.

Popper, K, (1972). *Objective knowledge: An evolutionary approach.* Oxford: Clarendon Press.

Powers, W. T. (1973). *Behavior: The control of perception.* Chicago: Aldine.

Powers, W. T. (1989). *Living control systems: Selected papers.* Gravel Switch, KY: CSG Books.

Robertson, R. J. & Powers, W. T. (Eds.). (1990). *Introduction to modern psychology: The control theory view.* Gravel Switch, KY: CSG Books.

Williams, W. D. (1989). Making it clearer. *Continuing the conversation: A newsletter of ideas in cybernetics,* 9-10.

Williams, W. D. (1990). The Giffen effect: A note on economic purposes. *American Behavioral Scientist, 34(1),* 106-109.

[1995]
Purpose, Context, and Synthesis: Can We Avoid Relativism?

This volume[1] is particularly welcome in this day and age. It appears in a context of postmodernism, deconstructionism, and poststructuralism, at a time when we are confronted with a plethora of claims concerning different ways of knowing, within a milieu in which the most important issue seems to be whether or not everyone's voice is heard. It sometimes seems as if reasoning, justification, validity, evidence, and claims of better and worse are relics of a bygone age. From individual relationships to talk shows to politics, everyone's opinion appears to be as good as everyone else's on almost any matter whatsoever. The very ideas of evaluation and the logic of evaluation would seem to be suspect. Truth and goodness seem to many to be relative to one's race, class, gender, and point of view.

However, the issue is not so simple as rejecting the relativists and returning to the good old days of revealed truth and positivistic science. Several of the most influential scholars in evaluation, including those in this volume, have been among the leaders in showing the limitations of classical, absolutist theories of truth, reasoning, and evaluation. They have expounded on the extent to which evaluation is relative to the purposes of the stakeholders, is context dependent, can yield different results using different kinds of evaluation, and is different within different ways of knowing. These are precisely the kinds of results pounced on by the relativists as undermining the possibility of using reasoning to reach warranted evaluative conclusions. Yet I believe that the authors remain committed to the legitimacy of evaluative reasoning.

So the present volume comes at a critical time as it tries to steer a course between the Scylla of absolutism and the Charybdis of relativism. What kinds of logic and reasoning might allow us to make warranted evaluative judgments, at least of better and worse, but that nonetheless appear to require the following features?

---

1 For context of this chapter, see About This Volume, page 352.

First published in: Deborah M. Fournier (ed.), *Reasoning in Evaluation: Inferential Links and Leaps*. A publication of the American Evaluation Association. Number 68, Winter 1995. San Francisco: Jossey-Bass.

- The judgments are dependent on the purposes, both individual and social, of the evaluator.
- The judgments are also dependent on the context in which the evaluation takes place; that is, the context determines the judgments that actually occur.
- Despite not being able to prespecify all of the evaluative criteria and how they will play out in a given context, we can make warranted evaluative syntheses, all things considered.

These three themes—purpose, context, and synthesis—pervade to a greater or lesser extent each of the four chapters. These themes are fundamental to a proper understanding of reasoning in evaluation, and each of the authors explores how a properly understood logic of evaluation can take them into account.

In the end, however, it appears that the authors share with the relativists a conception of context-dependent purposive human behavior that allows relativism to be seen as a plausible alternative. I will, therefore, add to their contributions by suggesting that the perceptual control theory view of human behavior gives an account of the interaction of purpose and context that renders warranted evaluative reasoning entirely plausible.

Purpose

The major contribution of the four main chapters is that, together, they implicitly show how at least some purposes can enter into evaluative reasoning in a perfectly straightforward, nonrelativistic way. Utilizing the notions of criterial definition and probative inference, Scriven demonstrates that many of the key concepts about which we wish to reason evaluatively have, as part of their criterial definitions, evaluative features. Then by making use of some very general, noncontroversial notions of merit we can infer to evaluative conclusions from the definitions of the items of interest along with certain factual premises. For example, if we know that the purpose of a cooking pot is to hold liquid, we can conclude that a leaky pot will not be nearly as good as one that does not leak.

However, purpose need not be considered as an essential part of the definition of a concept only in such "functional" cases. It also is part of the concepts of more overtly value-laden activities as well. As Smith describes in the evaluation of the *Iowa* explosion, the conclusion we reach will depend on whether we start with the concept of a legal liability to be adjudicated or a social scientific question to be resolved. The former concept relies heavily on expert opinion, whereas the latter is defined by generalizability and replicability. Thus Smith takes us beyond the individual purposes of clients, evaluators, and stakeholders to examine the societal purposes built into different games we have chosen to play in our social lives. He reminds us that human beings create certain kinds of social structures to carry out fundamental purposes shared across large groups of people, and as such those purposes serve as evaluative criteria within the concepts of these structures.

Fournier elaborates the discussion of purpose with her useful distinction between general and working logics. The purpose of general logic is to identify criteria, construct standards, devise measurements, and synthesize judgments. The purpose of a given working logic is to begin to flesh out these general features with the specific criteria, say, of a causal approach to the evaluation of program interventions. It would not be too much of a stretch to suggest that Fournier's working logic notion is a theoretical elaboration of Smith's ideas of societal games. As she describes the notion of working logic, in a connoisseurial approach to evaluation, for example, we would know from the concept of a connoisseur that legitimate evaluative conclusions from that approach will reflect the way it feels to an expert to be involved in the program.

House gives an extended example of the societal game of professors in a research university judging their colleagues for promotion and tenure. The purpose of such a game is to maintain the scholarly capacities of the university, and its working logic is expressed in the preparation of dossiers, letters from external referees, promotion and tenure committees, and the like.

The point is that the concepts we have, both of functional things and of social activities, carry a number of evaluative criteria within their very definitions. If we understand the concepts, then along with our experience in actually using those concepts to deal with the world we can, and often do, straightforwardly come to warranted evaluative conclusions. If we want to buy a cooking pot or a car, we compare what we see with the purposes of the pot or car and act accordingly. If we understand the purpose of the research university, we compare a candidate's dossier with our concept of a full professor and vote our conscience. If our purpose is to improve inner-city education, we compare a program's effects with our ideal of inner-city education and act accordingly. There simply is no logical gap between understanding a concept and being able to make evaluative judgments about instances of that concept in the real world in which we live our lives.

This is, of course, not to say that our concepts are immutable. In light of experience and other, higher-order goals, we can change and modify our concepts. With the advent of microwave cookery, certain metallic materials are no longer desirable in a cooking pot. As the social contract between society and the university is renegotiated more in the direction of undergraduate teaching and service, the concept of the full professor changes as well, as House notes in his chapter. Our concept of health care has gradually been evolving toward one that includes consideration for the quality of life as well as for its mere length. The process of reaching warranted judgments, however, remains the same. We compare our experience in the particular context with the evaluative criteria contained in the concept and the judgments follow.

Purpose and context

Typically, we view context as affecting the results of an evaluation when, roughly speaking, the same products or programs are being evaluated in different contexts with, perhaps, different results. For example, a Head Start program that works in one context is a disaster in another. In such a situation, we may well ask whether we can reach a warranted evaluation of whether Head Start is a good program or not. Indeed, this recurring phenomenon is the bane of much program evaluation.

Smith's example, on the other hand, emphasizes the fact that not only can the same societal game be played in different contexts, but sometimes different societal games with different purposes played in the same context can give different results. Fournier's notion of working logics likewise accents ways in which different purposes compel us to deal with different parts of the context. The working logics begin to fill in the major outlines of what features of the context will be relevant in any given evaluation such as causal, connoisseurial, and the like. In short, context and purpose are inextricably intertwined.

However, whether it is the same game played in different contexts or different games played in the same context, the key point is that what counts as contextually relevant is dependent on the purposes and goals of the evaluator, the other stakeholders, and the evaluation game being played. In the absence of these purposes, context is just another name for everything that there is—a notion much too all-encompassing to be of any use. Thus the fact that a given professor is a good spouse and parent is, for the most part, irrelevant to the promotion game, although it may be quite relevant to the good colleague and friend game. Furthermore, the candidate's friendships in the third grade are likely irrelevant to both (although if the candidate married a third-grade friend it could become relevant to the good colleague and friend game).

This last example illustrates an important feature of the ways in which purposes actually determine contexts. We tend to think that if we know the purposes of a given evaluation, we can derive criteria that will prespecify what parts of the context will be relevant, and, of course, to some extent, that is possible. However, as the examples of the four authors so abundantly demonstrate, whenever one thinks one has taken all of the contextual features into account, one can always construct a plausible story as to why something we forgot to consider is relevant after all.

This is an important theoretical point. Purposes do not determine contextually relevant features in a top-down way. Rather it is because of our ability to experience the world in light of our purposes that we find out what parts of the context are relevant. Actually performing the evaluation, gathering data, forming judgments, and considering alternative scenarios in the concrete setting of the evaluation all are compared to the desired purposes to determine the extent to which any of these activities actually seems to be contributing to the evaluation.

In this regard, House's claim that context limits possibilities seems to be exactly right. He illustrates this in numerous ways in discussing faculty evaluation. He comments on how letters of reference are important, but less so if from former advisors; how research is important, but a Nobel prize winner's research may outweigh all sorts of other criteria; and so on. The purposes do not preselect the relevant context; rather, the particular context triggers criteria in terms of which we are experiencing the situation.

Warranted syntheses

The central question of House's chapter is how a synthesis of various evaluations can be put together to reach a warranted evaluative judgment. He suggests that one often can, and, indeed, must put together the most coherent account possible of an evaluation. For House, coherence is the mark of a warranted synthesis. However, he emphasizes that the coherence depends on data and evidence and is always context bound. It is a coherent synthesis for the particular case and may or may not generalize easily, or at all, to other cases. His concrete examples are persuasive.

The evaluative descriptions of the syntheses seem to make sense, at least to most people. Nevertheless, the evaluative logic of the coherence criterion for House remains elusive. As he says, we can more often attain it than we can analyze it.

The other authors also give numerous examples of plausible syntheses of various evaluations being reached. But examples are not a theory of evaluative reasoning. If, as the authors seem to believe, evaluative judgments are relative to purposes, individual and social, how can we reach a warranted synthesis? If evaluative judgments are relative to context and it is not possible to prespecify the criteria that are relevant in the context, this is not just a complete capitulation to relativism? Despite the authors' examples, we still do not have any kind of account of how it is possible to account for purpose and context without falling into a radical relativism that simply says, "That's your story and I have mine."

It would be logically possible to try to address the threat of such a radical relativism by considering alternative conceptions of logic, or different "ways of knowing," or similar stratagems that might alleviate the tension between allowing for the relevance of purpose and context, and synthesis and the desire to reach warranted evaluative syntheses. There is, however, another alternative. The problem may lie not with our traditional conceptions of logic but with our traditional conceptions of human action that seem to treat purpose as at best a convenient fiction and at worst a mystifying superstition regarding human activity. It may be that with a more adequate conception of purposive human action, the threat of radical relativism will dissipate without the invention of new, esoteric logics or ways of knowing.

Perceptual Control Theory

Interestingly, over the past twenty years or so, a small, still largely unknown, body of work has begun to emerge that promises to meet the challenges to the logic of evaluation posed by the fact that our evaluative judgments are relative to our purposes, and the fact that the context determines the judgments that occur. This conception of human behavior was given its most powerful formulation by W. T. Powers (1973) in his book *Behavior: The Control of Perception*. Most recently in education, it has been the subject of a spirited debate in the pages of *Educational Researcher* (Cziko, 1992a, 1992b; Amundson, Serlin, and Lehrer, 1992). Researchers from

Purpose, Context, and Synthesis: Can We Avoid Relativism? 347

a variety of other disciplines are also contributing to this body of research in general psychology (Powers, 1989, 1992; Robertson and Powers, 1990), experimental psychology (Bourbon, 1990; Hershberger, 1988; Marken, 1986, 1989, 1990, 1992), clinical psychology (Ford, 1989, 1994; Goldstein, 1990), education (Bohannon, Powers, and Schoepfle, 1974; Petrie, 1974, 1979, 1981, 1986), sociology (McClelland, 1994; McPhail, 1991; McPhail, Powers, and Tucker, 1992), ethology (Plooij, 1984; Plooij and van deRijt-Plooij, 1990), law (Gibbons, 1990), management (Forssell, 1993), and economics (Williams, 1989, 1990).

This new conception of human nature is called perceptual control theory, and as the title of Powers' book implies, it fundamentally turns our conception of human action on its head. Instead of viewing behavior as the outcome of stimuli or perceptions (as modified by cognition, emotions, or planning), perceptual control theory views behavior as the means by which a perceived state of affairs is brought to and maintained at a (frequently varying) reference or goal state.

Perceptual control theory makes use of the idea of the "circular causation" found in engineering control and servomechanism theory. Thermostats and cruise control systems are everyday examples of mechanical control systems that keep the perceptions of temperature or speed near the reference levels set for them. Such physical control systems were invented precisely because engineers wanted to create mechanical systems that behaved as we humans do as we go about the tasks of governing temperature, maintaining speed, tracking targets, and, in general, successfully pursuing our goals in a constantly changing environment. In doing so they created a theory that is much more amenable to modeling actual human behavior than the stimulus-response, input-output, independent variable-dependent variable kind of theory created by the psychologists.

How does perceptual control theory meet the challenges to evaluative logic outlined above? This brief space does not allow a full accounting of the theory, but perhaps I can draw enough analogies with familiar mechanical control systems to pique the interest of the reader in this revolutionary approach to understanding human behavior.

I will start with purpose. A central function of perceptual control theory is to account for intentionality and purpose. The fundamental phenomenon of human action is that we constantly are able to achieve our purposes or ends with varying means in a continuously changing environment. Control systems are precisely what allows that to occur.

Consider the mechanical cruise control system. There are an indefinite number of factors that might keep a car from maintaining a certain speed: headwinds, crosswinds, hills, curves, poor-quality gasoline, and so on. If we tried to build a mechanical system that would be able to determine how any of these features might interfere with the desired speed (which itself might change from time to time during a trip) and also include the capacity for calculating just how much gas to deliver to the engine to overcome any of these disturbances, we might well conclude that we could not possibly build a cruise control system.

Similarly, if we try to specify our evaluation criteria completely before the fact and give appropriate weights to the various criteria for a good professor before we enter into the actual evaluation, we might well conclude that a completely warranted evaluation is not possible either. As House illustrates, we cannot possibly anticipate every nuance of context that might make a difference.

But the engineers did not try to anticipate every potential disturbance to maintaining speed. Rather, they built a mechanism that sensed the speed of the car, compared that speed to the desired one, and, if it was too slow, fed more gas to the engine and, if too fast, decreased the gas. The cruise control neither knows nor cares what factors in the environment cause the speed to depart from its desired level; it just compensates when it does. In short, it controls the perception of the speed of the car, keeping it very close to the desired level, and it does this in a circular causation kind of way in which the output affects the input at the same time as the input is being compared with the reference speed and the difference between the two is actuating the output.

Note, too, that this is just how the human driver without cruise control behaves. We do not check headwinds or hills, especially if they are slight, and compute how much to depress or let up on the accelerator. Rather we monitor the speedometer and no matter what the cause of a change in the speed we want to maintain, we depress or let up on the accelerator accordingly.

Similarly, the human evaluator monitors the situation in light of the evaluation criteria contained in the concept of what is being evaluated. The evaluator gathers data, learns about the stakeholders, considers alternative scenarios in the concrete setting of the evaluation, and uses whatever information comes from the actual case to form the appropriate evaluative judgment.

Context is simply the world that we experience. A part of the world becomes important to us if it constitutes a disturbance to some part of the experience we are trying to control. We recognize the disturbance as a difference in what we perceive compared to what we want to perceive, and in a well-functioning control system our actions tend to move our perceptions in the directions we want to see. Control systems do not need to have a prespecification of the context in which they operate; they only need to be able to sense changes in the perceptual variables they are trying to control, whatever causes those changes. It is through conceiving purposeful action as the operation of a control system, controlling for seeing the world as its purpose defines it, that we understand how purpose defines the context but cannot predetermine what will be relevant.

Context limits the possibilities. Within the range of those features of the context that might actually affect the evaluation, through collecting data, sifting evidence, and generally becoming familiar with the particular case, the evaluator senses those features that affect the overall goal of judging how well the product or process or program comes up to the concept of that product, process, or program. Because, as Scriven has shown, the evaluative criteria are, at least in part, built into the concept of what is being evaluated, comparing the actual situation to that concept automatically results in a judgment of how closely the situation matches the concept. It really is true that there are a number of concepts that we may not be able to define, but that we can recognize when we see. We do not need to account for the Nobel Prize winner in setting forth criteria for promotion, but if we see one, we recognize the achievement as relevant.

In an evaluation, we have some concept of what the ideal cooking pot or professor or program should look like. We go about the job of examining a particular example of that cooking pot or professor or program and, to the extent that what we see, wherever it comes from, is different from what we want to see, we are able to issue evaluative judgments. Then, given our typical beliefs about the way the world of policy and evaluations work, we hope our judgments will lead to changes in our perceptions of how the pots, professors, and programs are dealt with, changed, modified, or judged by others. (Of course, sometimes our evaluative judgments do not lead to the results we would like, causing a disturbance to another of our goals as evaluators and leading to other activities, such as writing papers about the relationship of evaluation to policy. But that is another story that could also be accounted for by perceptual control theory)

But someone may well object that other stories are precisely what distinguish real evaluations from mechanical cruise control systems. It may be interesting to note that the cruise control system does not have to prespecify the road conditions or weather or gasoline quality to maintain its set speed, but the real analogy to evaluation would lie in looking at situations in which it may be inappropriate to maintain the set speed at all. The cruise control system cannot help us there. It blindly maintains its set speed, even if, say, slowing and thickening traffic conditions call for a different speed. And, of course, this is correct. Analogies with mechanical control systems can go only so far, especially because most of the familiar ones are simple one-level control systems.

But consider the control system that is the human driver. The human driver is presumably maintaining the set speed in order to get somewhere. However, the driver presumably wants to get there safely as well, and all sorts of conditions may occur that would render a set speed inappropriate: traffic may slow; the speed limit may change; there may be an accident; a bridge may be flooded out; or the road may be slick from rain. There is no possible way that the human driver could prespecify all of the ways in which the trip at a given speed might become unsafe, but there is equally no question that when any of these things occur, we recognize them as a threat to our safe passage and change the speed accordingly.

We do some things in order to do others and those in order to pursue still higher-order purposes. The adaptable, intelligent person does not persist in behavior that does not serve higher-order goals but varies that behavior in an unpredictable, complex, and changing world in order to see the world as those higher-order goals specify it.

Perceptual control theory postulates the notion of a hierarchically arranged network of control systems such that the output of some of the higher levels changes the goals of the lower levels, bringing the whole into an equilibrium. This is what happens when we change the set speed of the cruise control in light of wanting to perceive ourselves as not crashing into the traffic ahead of us.

It is also in this way that we can begin to understand the idea of a coherent evaluative synthesis, even if we cannot prespecify how it would work out in particular cases. Consider House's promotion committee. Experienced professors have a concept of being a good professor in a research university without at all being able to define that concept in concrete terms. They sense the evidence regarding a given professor and can tell whether or not there is a deviation from their concept. Someone who receives poor undergraduate student ratings for teaching but has a record of successfully advising doctoral students can be seen as a good teacher of a different kind. An engineering professor who publishes few traditional journal articles but who is asked to serve on a national commission to set the standards for making concrete pipes is seen to be a good professor, exemplifying outstanding scholarship in a nontraditional way.

Our concept of a good professor plays the role of the desire to arrive where we are going safely in the driving example. Traditional refereed journal articles play a role analogous to the set speed in the cruise control system. We can recognize various ways in which refereed articles may contribute to research productivity, from top journals to prize-winning essays, without prespecifying all of these variations. At the same time, we can also see how maintaining an emphasis simply on refereed articles at all costs may lead us to a "traffic accident" with a professor who manifests outstanding research through an important consultative arrangement. We do not need to know all of the ways in which the context might affect our perception of the speed or the professor, but by comparing what we sense to our reference concepts, we know whether or not they fall short.

As House points out, we can even change our notion of a good professor in a research university as society presses us for better undergraduate teaching and more relevant research. The higher-order goal of those of us in the university maintaining a productive social contract with those who support us can vary the nature of lower-order goals we may have.

In an extremely important way, human beings are even more predictable than are physical events. Human beings are organized to attain consistent goals despite varying circumstances. An automobile mechanically programmed to drive around a given track will be less predictable in its path in the face of significant crosswinds than will the path of that same automobile in the hands of a human driver wanting to drive around the track. An evaluator applying predetermined criteria of evaluation will be less likely to reach a warranted synthesis than one alert to the unpredictable nuances of the actual case.

Thus there is nothing deficient in a logic of evaluation that does not allow us to specify in advance all the evaluative criteria and how different contexts might be judged in light of those criteria. Indeed, that is just what human beings, conceived as behaving so as to control their perceptions, do all the time. It is the only way in which we can make sense of the human capacity to achieve consistent results in a constantly changing environment.

We need not fear for the validity of evaluative syntheses that recognize the relativity to the purpose of the evaluation and to the context. By sensing the various nuances of the context, we are in effect comparing the actual context with our concept of that which we are evaluating. Because control systems control perceptions, not behavior, we need not know in advance what the details of the context might be. Once we have compared the concrete situation to our concept, however, we can describe the extent to which the situation meets or fails to meet our concept. In short, we can legitimately make warranted evaluative judgments.

About This Volume

In her Editor's Notes, volume editor Deborah M. Fournier explains how the first four chapters in this volume relate to the commentary by Hugh Petrie, the chapter reproduced above.

In Chapter One, *The Influence of Societal Games on the Methodology of Evaluative Inquiry*, Nick L. Smith orients the reader in thinking about reasoning at a broad level. He argues that good evaluative reasoning depends on the broader social enterprise or "game" within which an evaluation is being conducted, because the game into which the evaluator will be entering when conducting an evaluation influences the development and justification of claims. The game defines the purpose of the inquiry, the kinds of the phenomena being examined, the outcomes sought, the procedural rules, and ethics.

Comparing two inquiry games, criminal justice and social science, Smith illustrates how the justification of conclusions varies even though the same evidence and phenomenon are under investigation. This case example shows the usefulness of the game metaphor in helping evaluators determine appropriate evaluative reasoning and the subsequent use of strategies.

In Chapter Two, *Establishing Evaluative Conclusions: A Distinction Between General and Working Logic*, I shift the discussion of evaluative reasoning to a more detailed level and offer the notions of a general and working logic to explain how evaluators reason to establish and legitimate claims made in evaluation. I contend that the general logic of evaluation overarches all the various approaches and models within evaluation. This specifies what it means to evaluate something, an activity logically distinct from, say, biomedical research. In Smith's terms, general logic specifies the game.

Subsumed under the general logic is a profusion of individual working logics. Working logic is the variation in detail in which the general logic is followed. It is specific to a particular approach in terms of the type of problem, phenomena, questions, and claims of interest to the evaluator. The concepts of general and working logic highlight some of the important aspects of reasoning that evaluators must consider in building strong evaluations.

In Chapter Three, *Putting Things Together Coherently: Logic and Justice*, Ernest R. House takes the general logic introduced in Chapter Two and focuses on one part of it, that of integrating all the data sources into a final judgment. This is a fundamental problem that faces all evaluators. House proposes a general approach to the difficulty in combining multiple facts, values, interests, needs, preferences, and measures into conclusions about the evaluand. He suggests that evaluators fit together all the available information into the most coherent account that results in an all-things-considered synthesis judgment. Critical to reaching the synthesis judgment is the notion of context because it constrains the options possible.

In Chapter Four, *The Logic of Evaluation and Evaluation Practice*, Michael Scriven summarizes some of the problems alluded to in part by the previous authors. He identifies eight reasoning problems that face evaluation today and examines two of these problems in more depth: the problem of how one can ever get from empirically supported premises to evaluative conclusions and the problem of when it is and is not possible to infer from evaluative conclusions about a program, for example, to a recommendation as to what should be done with the program. These are serious cracks and flaws in the very foundations of evaluation practice and theory As Scriven suggests, a better understanding of fundamental issues and terminology and an explicated theory of evaluation are greatly needed.

The remaining three chapters offer commentary on the ideas presented in the first four chapters from three perspectives: informal logic, philosophy, and practice.

...

From the perspective of philosophy, Hugh Petrie, in Chapter Six, *Purpose, Context, and Synthesis: Can We Avoid Relativism?*, draws attention to what he sees as three common threads running across all four authors' positions on reasoning in evaluation: purpose, context, and synthesis. He points out that if these themes are fundamental to an understanding of reasoning in evaluation, then the authors share with the relativists a conception of purposive contextually bound behavior that permits relativism to flourish and undermines the possibility of reaching warranted evaluative conclusions. Petrie argues, however, that assuming the perspective of perceptual control theory of human behavior suggests to evaluators how to reach conclusions legitimately even though the judgments are inevitably dependent on purpose and context. Thus, to account for purpose and context without falling into a radical relativism, he addresses the threat by reexamining the traditional conception of human action.

...

It is hoped that through these seven chapters, the reader will be able to think more clearly and critically about logical practice; to appreciate the central role of reasoning and its analysis in the successful practice of evaluation. The discussion illuminates ways in which reasoning is influenced and open to challenge and marks out the greatest hurdles. The many unanswered questions raised throughout the issue should serve as fertile ground in promoting further investigation and development into what it means to establish sound evaluative reasoning in day-to-day practice. ...

Deborah M. Fournier
Editor

References

Amundson, R., Serlin, R. C., and Lehrer, R. "On the Threats That Do *Not* Face Educational Research." *Educational Researcher,* 1992,21 (9), 19-24.

Bohannon, P., Powers, W. T., and Schoepfle, M. "Systems Conflict in the Learning Alliance." In L. J. Stiles (ed.), *Theories for Teaching.* New York: Dodd, Mead, 1974.

Bourbon, W. T. "Invitation to the Dance: Explaining the Variance When Control Systems Interact." *American Behavioral Scientist,* 1990,34 (1), 95-105.

Cziko, G. A. "Perceptual Control Theory: One Threat to Educational Research Not (Yet?) Faced by Amundson, Serlin, and Lehrer." *Educational Researcher,* 1992a, *21* (9), 25-27.

Cziko, G. A. "Purposeful Behavior as the Control of Perception: Implications for Educational Research." *Educational Researcher,* 1992b, *21* (9), 10-18.

Ford, E. E. *Freedom from Stress.* Scottsdale, Ariz.: Brandt, 1989.

Ford, E. E. *Discipline for Home and School.* Scottsdale, Ariz.: Brandt, 1994.

Forssell, D. C. "Perceptual Control: A New Management Insight." Engineering *Management Journal,* 1993,5 (4), 1-7.

Gibbons, H. *The Death of Jeffrey Stapleton: Exploring the Way Lawyers Think.* Concord, N.H.: Franklin Pierce Law Center, 1990.

Goldstein, D. M. "Clinical Applications of Control Theory." *American Behavioral Scientist,* 1990,34 (1), 110-116.

Hershberger, W. A. (ed.). *Volitional Action: Conation and Control.* Amsterdam: Elsevier, 1988.

McClelland, K. "Perceptual Control and Social Power." *Sociological Perspectives,* 1994,37, 461-496.

McPhail, C. *The Myth of the Madding Crowd.* New York: Aldine DeGruyter, 1991.

McPhail, C., Powers, W. T., and Tucker, C. W. "Simulating Individual and Collective Action in Temporary Gatherings." *Social Science Computer Review,* 1992,10 (1), 1-28.

Marken, R. S. "Perceptual Organization of Behavior: A Hierarchical Control Model of Coordinated Action." *Journal of Experimental Psychology: Human Perception and Performance,* 1986, 12, 267-276.

Marken, R. S. "Behavior in the First Degree." In W. A. Hershberger (ed.), *Volitional Action: Conation and Control.* Amsterdam: Elsevier, 1989.

Marken, R. S. "Purposeful Behavior: The Control Theory Approach." *American Behavioral Scientist,* 1990,34 (1), 6-13.

Marken, R. S. *Mindreadings: Experimental Studies of Purpose.* Gravel Switch, Ky.: CSG Books, 1992.

Petrie, H. G. "Action, Perception, and Education." *Educational Theory,* 1974, 24, 33-45.

Petrie, H. G. "Against 'Objective' Tests: A Note on the Epistemology Underlying Current Testing Dogma." In M. N. Ozer (ed.), *A Cybernetic Approach to the Assessment of Children: Toward a More Humane Use of Human Beings.* Boulder, Cole,: West view Press 1979.

Petrie, H. G. *The Dilemma of* Inquiry *and* Learning. Chicago: University of Chicago Press, 1981.

Petrie, H. G. "Testing for Critical Thinking." *Proceedings of the Philosophy of Education Society 1985,* 1986.

Plooij, F. X. *The Behavioral Development of Free-Living Chimpanzee Babies and Infants.* Norwood, N.J.: Ablex, 1984.

Plooij, F. X., and van deRijt-Plooij, H. H. "Developmental Transitions as Successive Reorganizations of a Control Hierarchy." *American Behavioral Scientist,* 1990, 34 (1), 67-80.

Powers, W. T. *Behavior: The Control of Perception.* Hawthorne, N.Y.: Aldine DeGruyter, 1973.

Powers, W. T. Living *Control Systems: Selected Papers.* Gravel Switch, Ky.: CSG Books, 1989.

Powers, W. T. Living *Control Systems II: Selected Papers.* Gravel Switch, Ky.: CSG Books, 1992.

Robertson, R. J., and Powers, W. T. (eds.). *Introduction to Modern Psychology: The Control Theory View.* Gravel Switch, Ky.: CSG Books, 1990.

Williams, W. D. "Making It Clearer." *Continuing the Conversation: A Newsletter of Ideas* in *Cybernetics,* 1989, 18, 9-10.

Williams, W. D. "The Giffen Effect: A Note on Economic Purposes." *American Behavioral Scientist,* 1990, 34 (1), 106-109.

www.ingramcontent.com/pod-product-compliance
Lightning Source LLC
Chambersburg PA
CBHW050547160426
43199CB00015B/2566